Differential Geometry and Mathematical Physics

MATHEMATICAL PHYSICS STUDIES

A SUPPLEMENTARY SERIES TO
LETTERS IN MATHEMATICAL PHYSICS

VOLUME 3

Differential Geometry and Mathematical Physics

Lectures given at the Meetings of the Belgian Contact Group on Differential Geometry held at Liège, May 2-3, 1980 and at Leuven, February 6-8, 19

Edited by

M. CAHEN
Université Libre de Bruxelles, Belgium

M. DE WILDE
Université de Liège, Belgium

L. LEMAIRE
Université Libre de Bruxelles, Belgium

and

L. VANHECKE
Katholieke Universiteit Leuven, Belgium

D. Reidel Publishing Company

Dordrecht : Holland / Boston : U.S.A. / London : England

Library of Congress Cataloging in Publication Data
Main entry under title:

Differential geometry and mathematical physics.

(Mathematical physics studies ; v. 3)
Includes index.
1. Geometry, Differential–Addresses, essays, lectures.
2. Mathematical physics–Addresses, essays, lectures. I. Cahen, M. (Michel), 1935- . II. Belgian Contact Group on Differential Geometry. III. Series.
QC20.7.D52D538 1982 516.3'6 82-20535
ISBN-13: 978-90-277-1508-1 e-ISBN-13: 978-94-009-7022-9
DOI: 10.1007/978-94-009-7022-9

Published by D. Reidel Publishing Company
P.O. Box 17, 3300 AA Dordrecht, Holland

Sold and distributed in the U.S.A. and Canada
by Kluwer Boston Inc.,
190 Old Derby Street, Hingham, MA 02043, U.S.A.

In all other countries, sold and distributed
by Kluwer Academic Publishers Group,
P.O. Box 322, 3300 AH Dordrecht, Holland

D. Reidel Publishing Company is a member of the Kluwer Group

TABLE OF CONTENTS

PREFACE

This volume contains the text of the lectures which were given at the Differential Geometry Meeting held at Liège in 1980 and at the Differential Geometry Meeting held at Leuven in 1981. The first of these meetings was more orientated toward mathematical physics; the second has a stronger flavour of analysis.

The Editors are pleased to thank the lectures who contributed scientifically to these two meetings.

They are also grateful to Professor M. Flato who has encouraged publication of these contributions in the Mathematical Physics Studies Series.

We also thank the F.N.R.S. who supported financially the Contact group in differential geometry.

The Université de Liège and the Katholieke Universiteit Leuven which have given us a warm hospitality have contributed to the success of these meetings. We express our gratitude.

The Editors.

Lectures given at the
Meeting of the Belgian Contact Group
on Differential Geometry
held at Liège, May 2-3, 1980

SIMULTANEOUS DEFORMATIONS OF A LIE ALGEBRA AND ITS MODULES

D. Arnal

University of Dijon

INTRODUCTION

We expose here some results which are obtained by a team at the University of Dijon. This team included Jean-Claude Cortet, Georges Pinczon and myself.

We studied the question of the simultaneous deformation for a Lie algebra and one of its finite dimensional module. In fact, during the years 65-70, several authors considered these problems apart and the cohomological tools fastend to these questions were defined only for each aspect of these notions of deformations. Nevertheless the general problem is quite natural and moreover it happens when we consider examples of application. Indeed, Lie groups and algebras appear in physical problems uniquely in a particular representation. Thus, if we ask for deformations either as an "inversion" of contraction, or in the context of a perturbation theory, it is a general problem of simultaneous deformation of structures and representations of structures.

In 1966, Nijenhuis and Richardson (1) proposed a notion of graded Lie algebra associated to questions of deformations of structure or of morphism. Their description of these problems is interesting especially because the equations for each question are

M. Cahen et al. (eds.), Differential Geometry and Mathematical Physics, 3–15.

exactly the same. In the other hand, they describe the set of solutions in a neighbourhood of a point with the groups of cohomology, avoiding the introduction of the formal power series of deformation of Gerstenhaber (2).

Among the numerous authors which considered these questions, we mention also Hochschild and Serre (3), Levy-Nahas (4), Chevalley and Eilenberg (5).

Starting with the graded Lie algebras studied by Nijenhuis and Richardson, our goal was to define a graded Lie algebra associated to the general problem of deformations, in such a manner that a deformation was a solution of an equation formally identical with the previously considered equations. Then we determine the corresponding cohomological groups from those of partial problems thanks to a spectral sequence.

In this lecture, we begin by a short review of the theory of Nijenhuis and Richardson, with the object of defining notations and basic theorems. Then we give the construction of the graded Lie algebra that we consider, we establish the existence of the spectral sequence. Finally, we discuss completely an example of utilization of our methods.

I. NOTION OF GRADED LIE ALGEBRAS. DEFINITIONS (1).

In the following, K is a field with characteristic 0. A *graded algebra* A is a K-algebra such that :

$$A = \bigoplus_{\alpha \in \mathbb{Z}} A^{\alpha}$$

where each A^{α} is a subspace of A and where $A^{\alpha}.A^{\beta} \subset A^{\alpha+\beta}$ $\forall \alpha, \beta \in \mathbb{Z}$.

A is *anticommutative* if :

$$ab = -(-1)^{\alpha\beta}\, b\, a \qquad \forall\, a \in A^{\alpha} \qquad \forall\, b \in A^{\beta} \quad \forall\, \alpha, \beta.$$

A *graded Lie algebra* is a graded algebra with a product denoted by [,] anticommutative and satisfying the Jacobi identity :

$$(-1)^{\alpha\gamma}[[a,b],c] + (-1)^{\beta\alpha}[[b,c],a] + (-1)^{\gamma\beta}[[c,a],b] = 0 \quad \forall a \in A^{\alpha}, \quad \forall\, b \in A^{\beta}, \quad \forall c \in A^{\gamma}$$

Let A be a graded algebra, we define a new product on A by :

$$[a,b] = ab - (-1)^{\alpha\beta}\, ba \qquad \forall a \in A^{\alpha}\ \forall\, b \in A^{\beta}.$$

Then A becomes a graded Lie algebra in each of the following cases:

- A is associative or
- in A, the relation :

$$a(bc) - (ab)c = (-1)^{\beta\gamma}\,(a(cb) - (ac)b) \qquad \forall a \in A^{\alpha}, \forall b \in A^{\beta}, \forall c \in A^{\gamma}$$

holds.

If A is a graded algebra, a linear application D from A to A is an *homogeneous derivation of degree* α if :

$$D(A^{\beta}) \subset A^{\alpha+\beta} \text{ and } D(bc) = (Db)c + (-1)^{\alpha\beta}\, b(Dc) \quad \forall b \in A^{\beta},\ c \in A.$$

If A is a graded Lie algebra and $a \in A^{\alpha}$, the map ad a defined by :

$$\text{ad } a\ (b) = [a,b]$$

is a derivation of degree α.

The space D(A) of derivations of A is an associative graded algebra, we can thus define on it a structure of graded Lie algebra.

If D is an homogenous derivation of A such that $D^2 = 0$, we define the graded subalgebras :

$Z(A,D) = \text{Ker } D$ (its elements are cocycles)

and $B(A,D) = \text{Im } D$ (its elements are coboundaries).

B(A,D) is a graded ideal of Z(A,D), the quotient graded algebra :

$$H(A,D) = Z(A,D)/B(A,D)$$

is the cohomology associated to D.

II. SOME EXAMPLES IN THE THEORY OF DEFORMATIONS

a) Grassmann algebra

Let V be a finite dimensional vector space on K. The space $\wedge V$ of alternating multilinear forms with the product defined by :

$$\forall f \in \wedge^{k}V,\ g \in \wedge^{\ell}V : (f \wedge g)(v_1,\ldots,v_{k+\ell}) = \sum_{\substack{\sigma \in S_{k+1} \\ \sigma(1)<\ldots<\sigma(k) \\ \sigma(k+1)<\ldots<\sigma(k+\ell)}} f(v_{\sigma(1)},\ldots v_{\sigma(k)})\, g(v_{\sigma(k+1)},\ldots,v_{\sigma(k+\ell)})$$

is a graded associative algebra, Its associated Lie algebra is trivial :

$$[f,g] = 0 \quad \forall f,g$$

b) Deformations of morphisms

Let G and H be two finite dimensional Lie algebras on K, the space E of multi-linear alternating maps from G to H ($E = (\Lambda G) \otimes H$) with the gradation

$$E^n = (\Lambda^n G) \otimes H$$

and the bracket:

$$\forall \Phi = \omega \otimes u \, , \Psi = \pi \otimes v \quad [\Phi,\Psi] = (\omega \wedge \pi) \otimes [u,v]$$

is a graded Lie algebra. Its structure does not depend of the structure of G . Nevertheless, the G-cohomology is usually defined by an operator :

$$d : \Lambda G \to \Lambda G$$

defined by :

$$f \in \Lambda^n G \ (df)(x_1,\dots,x_{n+\ell}) = \sum_{i<j} (-1)^{i+j} f([x_i,x_j],x_1,\dots,\hat{x}_i,\dots,\hat{x}_j,\dots,x_{n+\ell})$$

Thus, the application $D = d \otimes 1$ is a derivation of degree 1 of E, $D^2 = 0$ and D depends uniquely of the structure of G.

Now, let ρ be a map from G to H. ρ is a morphism if and only if

$$\rho([x,y]) = [\rho(x) \, , \rho(y)] \qquad \forall x,y \in G.$$

In E, we can write this equation in the following form:

$$\rho \in E^1 \text{ and } D\rho + \frac{1}{2}[\rho,\rho] = 0 \qquad \text{(morphism equation)} .$$

Indeed:

$$(D\rho)(x_1,x_2) = -\rho([x_1,x_2])$$

$$[\rho,\rho](x_1,x_2) = [\rho(x_1), \rho(x_2)] - [\rho(x_2), \rho(x_1)] = 2[\rho(x_1),\rho(x_2)].$$

Let us put $D_\rho = D + \text{ad}\,\rho$. D_ρ is a derivation of E, thanks to the morphism equation and the Jacobi identity, we prove that

$$D_\rho^2 = 0 .$$

The complex $H(D_\rho,E)$ constructed with D_ρ is in fact the cohomology of Hochschild-Serre of the G-module H :

$$\forall \Phi \in E^n, (D_\rho\Phi)(x_1,\dots,x_{n+1}) = \sum_{i<j} (-1)^{i+j} \Phi([x_i,x_j],x_1,\dots,\hat{x}_i,\dots,\hat{x}_j,\dots,x_{n+1}) - \sum_{i=1}^{n+1} (-1)^i [\rho(x_i),\Phi(x_1,\dots,\hat{x}_i,\dots,x_{n+1})].$$

Finally, if u belongs to E^1, $\rho+u$ is a morphism if and only if it satisfies the morphism equation, i.e. if :

$$D_\rho(u) + \frac{1}{2}[u,u] = 0 \qquad \text{(deformation equation).}$$

c) Deformation of structures

Let G be a finite vector space on K, let us consider :

$$E = (\wedge G) \otimes G$$

graded by $E^n = (\wedge^{n+1} G) \otimes G$.

We define on E a structure of graded algebra, putting:

$\forall \Phi \in E^k$, $\Psi \in E^n (\Phi \circ \Psi)(x_1,\ldots,x_{n+k+1}) =$

$$\sum_{\substack{\sigma \in S_{n+k+1} \\ \sigma(1)<\ldots<\sigma(n+1) \\ \sigma(n+2)<\ldots<\sigma(n+k+1)}} \varepsilon(\sigma)\ \Phi(\Psi(x_1..x_{n+1}),x_{n+2},\ldots,x_{n+k+1}$$

The associativity does not hold on E, but the relation:

$\forall \Psi \in E^n, \theta \in E^k$: $\Phi\circ(\Psi\circ\theta) - (\Phi\circ\Psi)\circ\theta = (-1)^{nk}(\Phi\circ(\theta\circ\Psi)-(\Phi\circ\theta)\circ\Psi)$

holds. We can define a structure of graded Lie algebra on E. In particular, we have :

$$\forall\ \Phi \in E^1 : [\Phi,\Phi]\ (x_1,x_2,x_3) = 2\ S\ \Phi(\Phi(x_1,x_2),x_3)$$

where S means that we add for every circular permutation of (1,2,3). The structure equation will be in E :

$$\Phi \in E^1 \quad \text{and} \quad [\Phi,\Phi] = 0.$$

It is the same as the morphism equation with D = 0. Let us put $D_\Phi = \mathrm{ad}\Phi$. D_Φ is a derivation of E, $D_\Phi^2 = 0$, we associated to D a complex $H(D_\Phi,E)$ which is, up to order, the Hochschild-Serre cohomology for the Lie algebra (G,Φ) since :

$$\forall \Psi \in E^n\ D_\Phi\Psi(x_1,\ ,x_{n+2})=(-1)^n \sum_{i=1}^{n+2} (-1)^{i+1}[x_i,\Psi(x_1,\ ,\hat{x}_i,\ ,x_{n+1})]+$$

$$+ (-1)^n \sum_{i<j} (-1)^{i+j}\Psi([x_i,x_j],x_1,\ ,\hat{x}_i,\ ,\hat{x}_j,\ ,x_{n+2}).$$

Finally, if u belongs to E^1, $\Phi + u$ is a deformation of Φ if and only if $[\Phi + u,\ \Phi + u]$ vanishes, i.e. if and only if :

$$u \in E^1 \quad D_\Phi(u) + \frac{1}{2}[u,u] = 0 \qquad \text{(deformation equation)}$$

III. THEORY OF ANALYTICAL DEFORMATION IN E (1)

Now, the deformations of structures or morphisms are defined in an identical manner in a graded Lie algebra E. We shall give here the principal theorems of the theory with this formalism in the analytic case.

Puting $m = D + \rho$, we can write the morphism equation:

$$[m,m] = 0$$

Thus, we consider only this case. Let us choose a graded decomposition of E in the form :

$$E = Z(D_m,E) \oplus C = B(D_m,E) \oplus H(D_m,E) \oplus C = B \oplus H \oplus C.$$

Let M be the set of solutions of the equation $[m,m] = 0$ in E^1. Let us define the Lie group G of linear mapping from E to E of the form $e^{\text{ad } a}$, $a \in E^o$.

First we establish the :

<u>Lemma</u>. *Let* m *be an element of* M, *then the application*

$$F : C^o \times H^1 \times C^1 \to E^1$$

defined by :

$$F(a,h,u) = e^{\text{ad } a}(m + h + U)$$

is a local diffeomorphism in the neighbourhood of 0 *in* $C^o \times H^1 \times C^1$ *to* E^1 .

It is a direct consequence of the relation:

$$dF_{(o,o,o)}(a,h,u) = h + u - D_m(a).$$

We infer of that the following theorem.

<u>Rigidity theorem.</u> *If* H^1 *is null, then* m *is rigid : there exists a neighborhood of* m *in* M *which contains only elements of the orbit* Gm *of* m.

<u>Proof</u>. The map : $P : C^1 \to E^2$ defined by $P(u) = [m+u,m+u]$ is one-to-one in the neighbourhood of 0 in C^1, since

$$d P_o = 2 D_m.$$

The points of M which are in the neighborhood of 0 in E^1 defined by the lemma are the points :

$$e^{\text{ad } a}(m+u) \qquad \text{with} \qquad P(u) = 0 .$$

The injectivity of P means that u vanishes.

In order to study the obstructions to deform m, we establish:

Lemma. *There exists a neighborhood* V *of* m *in* E^1 *such that if* $x \in V$ *and satisfies :*

$$[x,x] \in C^2$$

then $[x,x] = 0$

Proof. The relations $[x,[x,x]] = 0$ holds for every x. If $x = m+u$, we have

$$(\text{ad } u + D_m)([x,x]) = 0$$

But the map ad $u + D_m$ is analytic in u, one to one on C^2 for $u = 0$; it is still one to one in a neighborhood of 0.

Theorem. *If* $H^2 = 0$ *, there exists an analytic mapping* Φ *, from a neighborhood* V *of* 0 *in* Z^1 *to* C^1 *such that in the neighborhood of* m*, the elements of* M *are the elements :*

$$m + z + \Phi(z) \qquad \text{with } z \in V.$$

Proof. Let Π_B be the projection on B^2 with kernel $H^2 + C^2$, let F be the function from $Z^1 \times C^1$ to B^2 defined by :

$$F(z,u) = D_m(u) + \frac{1}{2}\Pi_B([z+u,z+u]) = \Pi_B([m+z+u,m+z+u]).$$

Since

$$\left.\frac{\partial F}{\partial u}\right|_{o,o} = D_m$$

the implicit function theorem give us a neighborhood V and a map Φ such that :

$$z \in V \quad \text{and } F(z,u) = 0 \iff u = \Phi(z).$$

Moreover, in the neighborhood of m, the equation :

$$[m+z+u \; , \; m+z+u] = 0$$

is equivalent to :

$$\Pi_B([m+z+u \; , \; m+z+u]) = 0$$

Our conclusion follows immediately.

The set M of solutions to the deformation equation is now an analytic manifold in the neighborhood of m. The first group of cohomology allows us to determine the inequivalent deformations (in the sense of our first lemma).

If $H^2 \neq 0$, we can define the map Φ as supra. Let us introduce the map Ω from V to H^2 defined by :

$$\Omega(z) = \Pi_H(z+\Phi(z), z + \Phi(z)) ,$$

Π_H being the projection on H_2 with kernel $B^2 + C^2$. We obtain:

Theorem. 1) *In the neighborhood of* m, x *belongs to* M *if and only if* x *is on the form :*

$$x = m+z+\Phi(z) \quad \text{where } z \in V \quad \text{and} \quad \Omega(z) = 0$$

2) *Let* $\mathcal{H} = \{m+h+\Phi(h)$ *where* $h \in V \cap H^1$ *and* $\Omega(h) = 0\}$, *the map:*

$$\Psi : \mathcal{H} \times C^o \to M \text{ defined by : } \Psi(k,a) = e^{ad\, a}\, k$$

is a local analytic homeomorphism.

Now, M is no more an analytic manifold in the neighborhood of m, it is only the set of zeroes of an analytic fonction. $\mathcal{H}$ is a locally complete family of solutions in the sense of the theorem.

IV. GRADED LIE ALGEBRA ASSOCIATED TO SIMULTANEOUS DEFORMATIONS

Now, let us consider a Lie algebra G and one of its modules V. We constructed two graded Lie algebras :

$$E_1 = \wedge G \otimes \mathcal{L}(V) \text{ and } E_2 = \wedge G \otimes G$$

respectively associated to morphism and structure deformations problems. Let m be an element of E_2^n, for each ρ of E_1^p, the formula:

$$D_m(\rho)(x_1,\ldots,x_{n+p}) = (-1)^m \sum_{\substack{\sigma \in S_{n+p} \\ \sigma(1)<\ldots<\sigma(n+1) \\ \sigma(n+2)<\ldots<\sigma(n+p)}} \rho(m(x_{\sigma(1)},\ldots,x_{\sigma(n+1)}),x_{\sigma(n+2)},\ldots,x_{\sigma(n+p)})$$

defines an element of E_1. An easy calculation shows :

Lemma. *The map* $m \mapsto D_m$ *is a graded Lie algebra morphism from* E_2 *to* E_1.

We are thus able to define a new graded Lie algebra $E_o = E_1 + E_2$ as a semi-direct product of E_1 and E_2:

if $\rho + m \in E_o^n = E_1^n + E_2^n$ and $\rho' + m' \in E_o^{n'}$:

$$[\rho + m, \rho' + m']_o = [\rho,\rho']_1 + D_m(\rho') - (-1)^{nn'} D_{m'}(\rho) + [m,m']_2 .$$

A point $m_o + \rho_o$ of E_o is a Lie structure and a morphism of this structure if and only if :

$$[m_o,m_o] = 0 \quad (\text{in } E_2) \text{ and } D\rho_o + \frac{1}{2}[\rho_o,\rho_o] = 0 \quad (\text{in } E_1).$$

But the derivation D in this last equation is ad m_o, computed in E_o. Then these equations can be written as :

$$2[m_o,\rho_o] + [\rho_o,\rho_o] = 0 \quad \text{and} \quad [m_o,m_o] = 0$$

Or equivalent as a structure equation in E_o:

$$[m_o + \rho_o , m_o + \rho_o] = 0 .$$

The deformation equation follows immediately : $m_o + \rho_o + m + \rho$ is a deformation of $m_o + \rho_o$ if and only if :

$$[m_o + \rho_o, m + \rho] + \frac{1}{2}[m + \rho, m + \rho] = 0 .$$

or if we prefer :

$$[m_o+m] + \frac{1}{2}[m,m] = 0 \text{ in } E_2 \text{ and } [m_o + \rho_o,\rho] + [\rho,m] + \frac{1}{2}[\rho,\rho]=0 \text{ in} E_1$$

We can apply all the previous results to the global problem. In particular, we introduce the cohomology groups H_o^i defined by the derivation $\mathrm{ad}(m_o + \rho_o)$. These groups give us the solutions and obstructions for solutions in our problem. We shall now determine these groups.

V. EXACT SEQUENCES OF COHOMOLOGY

Let us consider in E_1 the cohomology defined by $\mathrm{ad}(m_o+\rho_o)\big|_{E_1}$ and in E_2 the cohomology defined by $\mathrm{ad}\, m_o\big|_{E_2}$. These two cohomologies are those introduced in §2. We denote by Z_o^i (resp. Z_1^i, Z_2^i) the kernels of coboundary operators in E_o^i (resp. E_1^i, E_2^i) and B_o^i (resp. B_1^i, B_2^i) their ranges.

Let j be the canonical injection from E_1^i to E_o^i and π the projection map from E_o^i to E_2^i with kernel E_1^i. The sequence :

$$0 \to Z_1^i \overset{j}{\to} Z_o^o \overset{\pi}{\to} Z_2^i$$

is exact. Let θ_1 be the application from E_2^i to E_1^{i+1} defined by :

$$\theta_1(m) = [\rho_o,m]$$

if $m \in Z_2^i$, then $\theta_1(m)$ belongs to Z_1^{i+1}:

$$[\rho_o+m_o,[\rho_o,m]] = (ad(\rho_o+m_o))^2 (m) - ad(\rho_o+m_o)([m_o,m])=0.$$

In the quotients, we define thus a mapping θ from Z_2^i to H_1^{i+1} .

Lemma. *The sequence*

$$0 \to Z_1^i \overset{j}{\to} Z_o^i \overset{\pi}{\to} Z_2^i \overset{\theta}{\to} H_1^{i+1}$$

is exact.

Proof. Indeed, if $m \in \pi(Z_o^i)$, there exists ρ in E_1^i, such that :

$$ad(\rho_o + m_o)(\rho + m) = 0$$

or :

$$[\rho_o + m_o,\rho] + [\rho_o,m] = 0 .$$

Thus $\theta_1(m)$ is a coboundary and conversely.

We can restrict that exact sequence to coboundaries.

Lemma. *The sequence*

$$0 \to B_1^i + \theta_1(Z_2^{i-1}) \overset{j}{\to} B_o^i \overset{\pi}{\to} B_2^i \overset{\theta}{\to} 0$$

is exact.

Proof. Let us suppose that ρ+m is a coboundary (in B_o^i), then $\pi(\rho+m)$ vanishes if and only if ρ is an element of B_o^i. We can write:

$$\rho = [\rho_o+m_o, \xi+u] = [\rho_o+m_o,\xi] + [\rho_o,u] + [m_o,u] .$$

Or :

$$[m_o,u] = 0 \quad \text{and} \quad \rho = [\rho_o+m_o,\xi] + [\rho_o,u] .$$

Thus u belongs to Z_2^{i-1} and ρ is in $j(B_1^i + \theta_1(Z_2^{i-1}))$. Finally if m is in B_2^i, we can write

$$m = [m_o,u] .$$

And :

$$ad(\rho_o+m_o) [\rho_o,u] = - ad(\rho_o+m_o)[m_o,u] + (ad(\rho_o+m_o))^2(u)$$
$$= - ad\ \rho_o(m) - (ad\ m_o)^2(u) = - \theta_1(m).$$

From that we deduce :

Theorem. *The sequence*

$$\ldots\ldots \overset{\theta}{\to} H_1^i \overset{j}{\to} H_o^i \overset{\pi}{\to} H_2^i \overset{\theta}{\to} H_1^{i+1} \overset{j}{\to} \ldots\ldots$$

is exact.

VI. AN EXAMPLE.

The former theorem reduces the general problem to the partial ones (i.e. to the computation of H_1^i and H_2^i) and to the determination of θ .

We shall conclude with a simple example showing the rôle of this application θ . First we prove :

Proposition. *If* ρ_o *is the adjoint representation and if* ρ_o *is faithful, then* θ *vanishes.*

Indeed if m belongs to E_2^i, we define $\hat{m}$ in E_1^{i+1} by :

$$\hat{m}(x_1,\ldots,x_i)(x) = m(x,x_1,\ldots,x_i).$$

With our hypothesis, we prove that :

$$\widehat{[m_o,m]} = -[\rho_o,m] + [\rho_o+m_o,\hat{m}] .$$

If m is in Z_2^i , $\theta_1(m)$ is the coboundary of $\hat{m}$, θ vanishes.

Let us consider now the complex Lie algebra $E(2)$, m_o is given on a basis (Z,T_1,T_2) of $E(2)$ by the relations :

$$m_o(Z,T_1) = T_2$$
$$m_o(Z,T_2) = -T_1$$
$$m_o(T_1,T_2) = 0$$

The cohomology groups for this algebra are :

$$H_2^{-1} = 0,\ H_2^o = \mathbb{C},\ H_2^1 = \mathbb{C}^2,\ H_2^2 = \mathbb{C},\ H_2^n = 0 \qquad \forall n > 2$$

We can deform the algebra $E(2)$ only in two directions : because there exists a non trivial obstruction, the only admissible cocycles are (modulo coboundaries) :

1) $m_1(Z,T_i) = 0, \quad m_1(T_1,T_2) = Z$
($m_o + \lambda m_1$ is the algebra sl(2)) and

2) $m_2(Z,T_1) = m_2(T_1,T_2) = 0,\ m_2(Z,T_2) = T_2$
($m_o + \lambda\, m_2$ is a solvable, three dimensional Lie algebra G_λ).

The set $\mathcal{H}$ defined in §2 is, here a cone in $\mathbb{C}^2 = H_2^1$.

If ρ_o is the adjoint representation of m_o, the corresponding cohomology groups are :

$$H_1^{-1} = 0,\ H_1^o = \mathbb{C},\ H_1^1 = \mathbb{C},\ H_1^2 = \mathbb{C},\ H_1^3 = \mathbb{C},\ H_1^n = 0 \quad \forall n > 3 .$$

There is no obstruction to deform ρ_o, $\mathcal{H}$ is H_1^1. But ρ_o is faithful, thus θ vanishes, we find for H_o^i :

$$H_o^o = \mathbb{C}^2,\ H_o^1 = \mathbb{C}^3,\ H_o^2 = \mathbb{C}^2,\ H_o^3 = \mathbb{C},\ H_o^n = 0 \qquad \forall\, n > 3.$$

It remains an obstruction to deform : if the π projection of our deformation $\rho+m$ is $\lambda\, m_1$, ρ is determined by λ. $\mathcal{H}$ is in $\mathbb{C}^3 = H_o^1$ a cone, the product of a plane with a line.

Instead of ρ_o, let us consider the representation ρ'_o of m_o defined by :

$$\rho'_o(T_i) = \text{ad}\ T_i,\ \rho'_o(Z) = \text{Id} + \text{ad}\ Z\ .$$

Since the coboundary operators $\text{ad}(\rho_o+m_o)\big|_{E_1}$ and $\text{ad}(\rho'_o+m_o)\big|_{E_1}$ are the same, we find the same cohomology groups H_1^i, H_2^i and the same results for partial problems. However, the maps θ do not vanish. In fact, we find :

$$0 \to \mathbb{C} = H_2^o \overset{\theta}{\underset{\sim}{\to}} H_1^1 = \mathbb{C} \to 0$$

$$\mathbb{C}^2 = H_2^1 \overset{\theta}{\to} H_1^2 = \mathbb{C} \to 0$$

$$0 \to \mathbb{C} = H_2^2 \overset{\theta}{\underset{\sim}{\to}} H_1^3 = \mathbb{C} \to 0\ .$$

We deduce H_o^i of that :

$$H_o^o = \mathbb{C}^2,\ H_o^1 = \mathbb{C}^2,\ H_o^2 = 0,\ H_o^n = 0 \quad \forall\, n > 2$$

The first cohomology group dimensionality is lower as for ρ_o but the obstruction to deform disappears. The point $m_o + \rho'_o$ is a regular one in the set of solutions.

In fact, we can see that Tr $\rho'_o(Z)$ being non null, the simultaneous deformation to sl(2) and a sl(2)-module is now impossible.

REFERENCES

(1) A. Nijenhuis and R. Richardson : "Cohomology and deformations in graded Lie algebras". Bull. Amer. Math. Soc. 72(1966) 1.29.

(2) M. Gerstenhaber : "On the deformations of rings and algebras". Ann. of Math. 79 (1964) pp. 59-104.

(3) G. Hochschild and J.P. Serre : "Cohomology of Lie algebras" Ann. of Math. 57 (1958) pp. 591-603.

(4) M. Levy-Nahas : "Deformation and contraction of Lie algebras" J.M.P. vol. 8 n°6 (1967) pp. 1211-1223.

(5) C. Chevalley and S. Eilenberg : "Cohomology theory of Lie groups and Lie algebras" Trans. Amer. Math. Soc. vol. 63 (1948) pp. 85-124.

HOW TO CONNECT EXPONENTIAL INTEGRALS TO OPTIMAL PROCESS ?

E. Combet

Université de Lyon, France

0. INTRODUCTION

One of the most interesting problems in equilibrium statistical mechanics is the problem of critical exponents which appear in the neighbourhood of the critical point of the phase diagram :

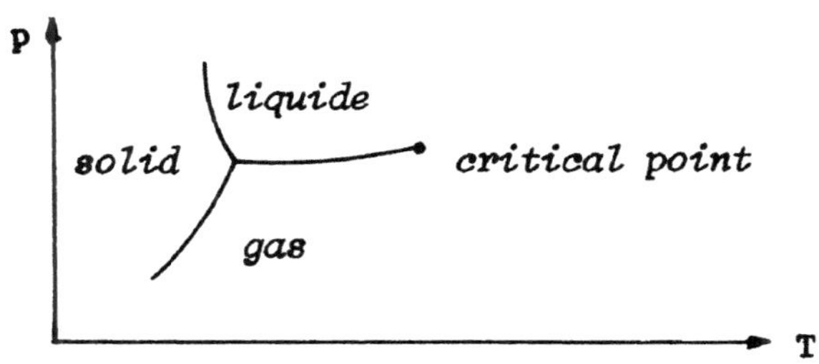

These numbers set a very intriguing problem : the critical exponents are apparently universal. Presumably, this means that there are only a few parameters and a few physical general principles which are relevant near the critical point (see C.J. THOMPSON (1), ch. 4).

In a first step, we can say that critical phenomena are physical systems where we find : exponential integrals (Gibbs ensembles, Wiener integrals), critical exponents (with large stability and universality properties) and fundamental structures which are present in physical phenomena (symplectic structure and

M. Cahen et al. (eds.), Differential Geometry and Mathematical Physics, 17–25.

optimal conditions).

In this talk, I shall consider a *geometrical system* which is *similar* to this physical situation (but I do not know how to return from this geometrical system to the physical ones !).

In the first part I shall study the lagrangian properties oi exponentials : this is quite parallel to J.J. DUISTERMAAT (2) but with a supplement on monodromy and Newton's diagram. In the second part I shall consider the lagrangian properties of optimal problems.

Results given here are detailed in my paper (3).

I. EXPONENTIAL INTEGRALS AND CRITICAL EXPONENTS.

We consider the exponential integral :

$$I(\omega) = \int e^{-\omega\phi(\theta)} f(\theta)d\theta$$

with a smooth phase function $\phi \in C^{\infty}(\mathbb{R}^N,\mathbb{R})$ and a smooth amplitude $f \in \mathcal{D}(\mathbb{R}^N)$.

Here ϕ is positive : $\phi \geqslant 0$ and $\phi(0) = 0$; therefore 0 is a critical point of ϕ.

1. Existence of critical exponents

It is required to find the asymptotic behavior of $I(\omega)$ as $\omega \to +\infty$.

Let us suppose ϕ *is analytic* : $\mathbb{R}^N \to \mathbb{R}$

Then by a stratification theorem of Whitney, $\phi^{-1}(t)$ is a regular hypersurface (near $0 \in \mathbb{R}^N$) for $t \in]0,t_o]$ with $t_o > 0$ enough small.

We write :

$$I(\omega) = \int_0^{+\infty} e^{-\omega t} \int_{\phi^{-1}(t)} f_t \cdot \eta_t$$

where f_t is the restriction : $f|_{\phi^{-1}(t)}$ and η_t the (N-1)-form : $\frac{d\theta}{d\phi}$ (t > 0).

By using Hironaka resolution theorem, the behavior of

$$\int_{\phi^{-1}(t)} f_t . \eta_t$$

as $t \to 0^+$ was given by P. JEANQUARTIER (4).

By Laplace method we then obtain the following development of $I(\omega)$ ((3), 2.10):

There exist an open neighbourhood U of $0 \in \mathbb{R}^N$ and an integer $q \in \mathbb{N}^\star$ such that for every $f \in \mathcal{D}(U)$, the following development is available :

$$I(\omega) \underset{\omega \to +\infty}{\sim} \sum_{\substack{j \in \mathbb{N} \\ k=0,1,\ldots,N-1}} a_{jk}\, \omega^{-\frac{j}{q}} (\mathrm{Log}\ \omega)^k .$$

Now we consider the critical exponents of I that is the maximal pair $(-\frac{j}{q}, k)$ which is effectively present in the development $I(\omega)$ for some function f and we look for the stability properties of these exponents.

2. Critical exponents and monodromy of $(\phi,0)$

I suppose again that ϕ is analytical : $\mathbb{R}^N \to \mathbb{R}$ but moreover *it has a complex extension :* $\mathbb{C}^N \to \mathbb{C}$ *which possesses an isolated singularity at* $0 \in \mathbb{C}^N$.

Then we can adapt to our exponential integrals the method given by B. MALGRANGE (5) for oscillatory integrals.

We return to $J(t) = \int_{\phi^{-1}(t)} f_t . \eta_t$.

We consider the Milnor fibration $\phi : X^\star \to T^\star$

$$X^\star = \{z \in \mathbb{C}^N \ ;\ \|z\| < \varepsilon,\ \phi(z) \in T^\star\}$$

$$T^\star = \{t \in \mathbb{C} \ ;\ t \neq 0,\ |t| < \eta\}$$

where ε, η are suitable; $X(t) = \phi^{-1}(t) \cap X^\star$ is the fiber of ϕ:

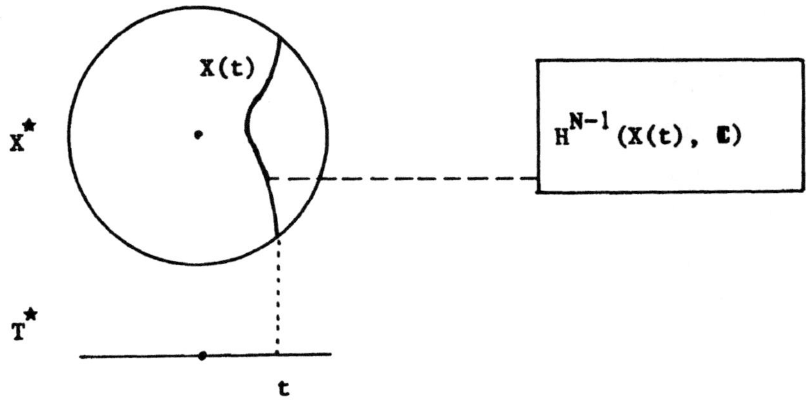

We get the complex domain by taking the Taylor expansion of f at any order and replacing J(t) by an integral :

$$\int_{\gamma(t)} \omega(t)$$

where $\gamma(t)$ is a continuous (N-1)-cycle in X(t) and ω an holomorphic section of the cohomological bundle $\bigcup_{T^*} H^{N-1}(X(t),\mathbb{C})$.

Let ∇ be the Gauss-Manin connection of the singularity $(\phi,0)$. Then the derivation formula :

$$D\left(\int_{\gamma(t)} \omega(t)\right) = \int_{\gamma(t)} \nabla\omega(t)$$

links the expansion of J at 0 to the singular point 0 of the fundamental system of ∇ and finally to the monodromy of $(\phi,0)$: the critical exponents of I appear in the eigenvalues of the monodromy ((3), 3.19).

3. Critical exponents and Newton's diagram

With generic suitable conditions, V.A. VASILEV result (6) shows that these critical exponents depend only on the Newton's diagram of $(\phi,0)$.

This diagram $\Gamma(\phi,0)$ is obtained by writing $\phi = \sum_{\nu} a_{\nu}\, \theta^{\nu}$, taking $K(\phi,0) = \{\nu \in \mathbb{N}^N \ ;\ a_{\nu} \neq 0\}$ and going to $\Gamma(\phi,0)$ by convex envelope:

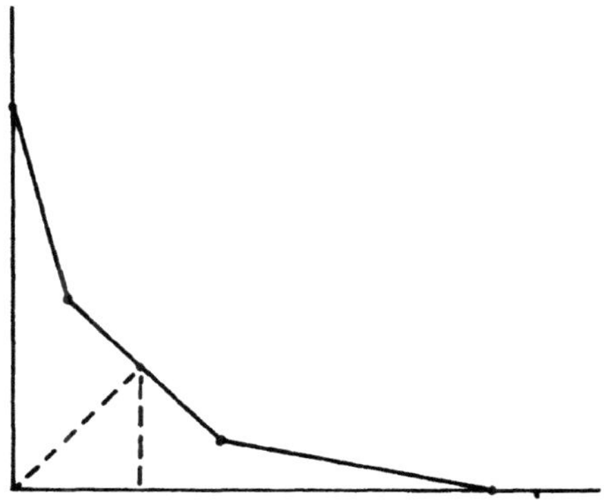

After V.A. VASILEV, these critical exponents are $(-\frac{1}{d(\Gamma)}, k(\Gamma))$ where $d(\Gamma)$ and $k(\Gamma)$ are classical numbers associated to Γ.

Thus *these critical exponents are very stable : they depend only on the smallest indices* ν *which are effectively present in the Taylor expansion of* ϕ .

4. Lagrangian properties of exponential integrals

In order to obtain a natural insertion of the theory in a physical background we have to study how to have exponential integrals in a symplectic space like a cotangent bundle $T^{\star}X$ with its fundamental 2-form $dp \wedge dq$.

This will be possible by considering Lagrangian submanifolds of $T^{\star}X$.

Indeed let Λ be a Lagrangian submanifold of $T^{\star}X$ (that is $\dim \Lambda = \dim X$ and $dp \wedge dq|\Lambda = 0$), then Λ is locally defined by non-degenerate phase functions $\phi \in C^{\infty}(X \times \mathbb{R}^N ; \mathbb{R})$:
$\Lambda = \{(x, \phi'_x(x,\theta)) ; \phi'_\theta(x,\theta) = 0\}$.

Here we adapt to positive phases ϕ the classical local Lagrangian analysis (see for example (2); also (3) for a positive ϕ). We consider the exponential integral :

$$(\star) \qquad \omega^{\frac{N}{2}} \int e^{-\omega\phi(x_0,\theta)} f(\theta)d\theta$$

at a point $\lambda_0 = (x, \phi'_x(x_0, \theta_0))$; then by Lagrangian calculus we can change the phase ϕ by changing suitably the amplitude f.

If the point λ_o is *stable* on Λ, we can prove that the critical exponents of (*) depend only on the singularity (Λ,λ_o) and the case of **interest** is when λ_o belongs to the caustic set $\Gamma(\Lambda)$ ((3), ch. 4).

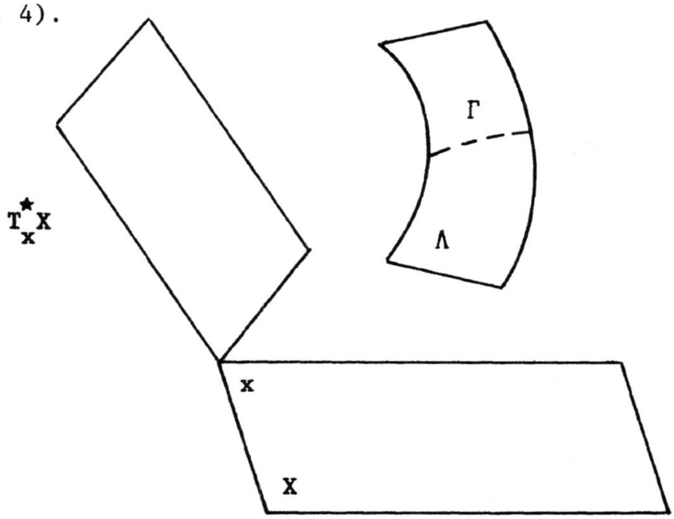

II. LAGRANGIAN PROPERTIES OF OPTIMAL PROCESSES

Now in order to insert the above result in a natural situation we have to study how to go to exponential integrals in an optimal problem. In this direction the first step is : how to obtain the Lagrangian properties of optimal processes in a space $\mathbb{R}^n$? For all results on optimal control we refer to W.H. FLEMING and R.W. RISHEL (7).

5. Deterministic process

Here we consider a system governed by a set of differential equations in $\mathbb{R}^n$:

$$\dot{\xi}(t) = f(\xi(t), \theta(t)), \qquad t \geq 0$$
$$\xi(0) = x \in \mathbb{R}^n ;$$

The function θ is a control : $\mathbb{R}^+ \to \mathbb{R}^n$.

The performance criterion is a function :

$$M(x,\theta) = \int_0^T L(\xi(t),\theta(t))dt + \Psi(\xi(T))$$

where T is the time when $\xi(T)$ belongs to a given hypersurface

$S \subset \mathbb{R}^n$ and Ψ a map : $S \to \mathbb{R}$:

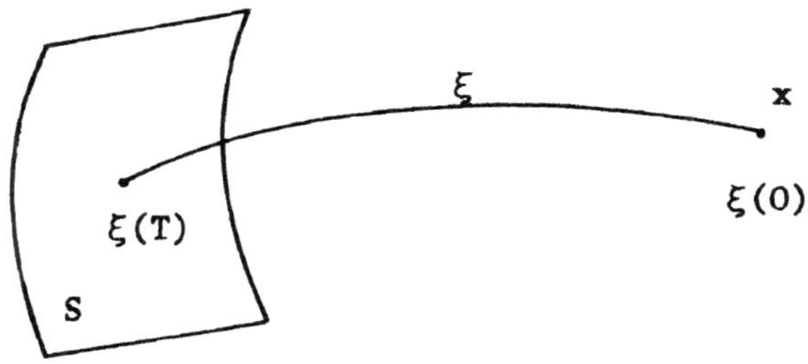

By the argument of dynamic programming, the value function:

$$V(x) = \inf_{\theta} M(x,\theta)$$

is a solution of the Cauchy problem :

$$H(x, \frac{\partial V}{\partial x}) = 0$$

$$V|S = \Psi$$

where H=0 is the Hamilton-Jacobi-Bellman equation :

$$H(x,p) = \inf_{u} [< f(x,u), p > + L(x,u)] .$$

By considering H as an Hamiltonian : $T^{\star} \mathbb{R}^n \to \mathbb{R}$, then a Lagrangian solution $\Lambda \subset H^{-1}(0)$ is generated by Hamiltonian flow associated to H via the symplectic form of $T^{\star}X$. There is a propagation of Cauchy data along bicharacteristic strips and the caustic set of Λ has a geometric origin. In a regular variational calculus, the Pontryagin-Weierstrass principle gives, (via Legendre transformation), a straightforward connection between variation calculus data and these Lagrangian solutions ((3), ch. 6) :

Focal points		Caustic set Γ
Broken extremals		Finite dim. phase variables
Index-function of M. Morse	Legendre $\longrightarrow$	Natural phase
Manifolds of paths		Infinite dim. phase variables
Action integral M		Generalized phase

In the Riemannian variational calculus, M is a generalized phase and exponential integrals are Wiener integrals : see (8).

6. Stochastic process

Here we consider a system governed by a set of stochastic differential equations indexed on a probability space Ω:

$$d\xi = f(\xi(t),\theta(t))dt + \sigma(\xi(t))dw, \quad t \geqslant 0$$
$$\xi(0) = x \in \mathbb{R}^n$$

The performance criterion is the expectation :

$$M(x,\theta) = E\left[\int_0^T L(\xi(t),\theta(t))dt + \Psi(\xi(T))\right]$$

where T is the exit time from a given bounded open set $G \subset \mathbb{R}^n$ and Ψ a map : $\partial G \to \mathbb{R}$.

By the argument of dynamic programming, the value function:

$$V(x) = \inf_{\theta} M(x,\theta)$$

is a solution of the Dirichlet problem :

$$\frac{1}{2} \sum_{i,j} a_{ij}(x) \frac{\partial^2 V}{\partial x_i \partial x_j} + H(x, \frac{\partial V}{\partial x}) = 0 \text{ (H.J.B. equation)}$$
$$V|\partial G = \Psi$$

where $a = \sigma^t \sigma$.

If the stochastic coefficient a is uniformly elliptic, V is a smooth map and the Lagrangian submanifold $\Lambda_V = \{(x, \frac{\partial V}{\partial x})\}$ has no caustic set. In the contrary case, if the matrix a is singular at x_0, it becomes possible for Λ_V to have a caustic point above x_0. Example (n=2) : consider the solution $r^{4/3}$ of the equation :

$$\frac{1}{2} r^2 \Delta V - \frac{9}{32} \|\text{grad } V\|^4 = 0.$$

In this case, Λ_V is obtained by diffusion of Dirichlet data, a caustic point (x_0,p_0) of Λ_V has a stochastic origin and $G - \{x_0\}$ must be taken in place of G in the estimation of T ; M will be a generalized phase function of Λ_V.

BIBLIOGRAPHY

(1) C.J. Thompson, Mathematical Statistical Mechanics, Princeton University Press. Princeton 1979.

(2) J.J. Duistermaat, Oscillatory Integrals, Lagrange Immersions...

Communications on P. and A. Mathematics, XXVII, 207-281 (1974).

(3) E. Combet, Intégrales exponentielles. Preprint 1981, Université de Lyon 1.

(4) P. Jeanquartier, Développement asymptotique de la distribution de Dirac, CRAS (série A), 271 (1970), 1159-1161.

(5) B. Malgrange, Intégrales asymplectiques et monodromie, Annales Sc. E.N.S., 4ème série 7, 1974, 405-430.

(6) V.A. Vasilev, Asymptotic exponential integrals ..., Functional An. Appl. 11 (1977), 163-172.

(7) W.H. Fleming, R.W. Rishel, Deterministic and Stochastic Optimal control., Springer-Verlag, 1975.

(8) B. Lascar, Le noyau de l'équation des ondes ..., Journal d'Analyse mathématique. 37 (1980), 1-31.

INVARIANCE OF MAXWELL'S EQUATIONS

S. Gutt

Aspirant du FNRS
Université Libre de Bruxelles CP 218

When special relativity is presented, one generally introduces Poincaré group as the group of invariance for electromagnetism. This definition is still rather unprecise; in particular the characterization of Poincaré's transformations is usually related to the wave equation in $\mathbb{R}^4$ and not directly to Maxwell's equations. We shall give here a possible definition of the invariance group of Maxwell's equations for free fields, in Minkowski space as well as in its conformal compactification Segal's model. Furthermore we give an explicit determination of all C^∞ solutions of Maxwell's equations in Segal's model.

The results have been obtained in collaboration with M. Cahen, we shall give here a summary of some of them. For more details we refer to (1) and (2).

Let (M,g) be a pseudo-riemannian manifold of dimension 4. A *Maxwell's field* on M is a C^∞ 2-form F on M such that

$$dF = \delta F = 0$$

where d is the exterior derivative on M and δ the codifferential on (M,g).

We define the *invariance group* A *of Maxwell's equations* as the group of C^∞ diffeomorphisms of M which preserve the space of

M. Cahen et al. (eds.), Differential Geometry and Mathematical Physics, 27–29.

Maxwell's fields. Explicitly :

$A = \{\phi : M \to M$ diffeo. such that if F is a C^∞ 2-form on M with

$dF = \delta F = 0$ then $d\,\phi^\star F = \delta\,\phi^\star F = 0\}$.

This definition led us to consider the commutation relations of the operators δ and $\mathcal{L}_X$ where X is a C^∞ vector field on M and $\mathcal{L}_X$ denotes the Lie derivative. One has :

1) *Let* X *be a vector field on* M *such that* $\delta\mathcal{L}_X\omega = \mathcal{L}_X\delta\omega$ *for all exact* p*-forms on* M *(where* p *is any fixed integer such that* $1 \leqslant p \leqslant \dim M$). *Then* X *is an infinitesimal isometry.*

2) *A vector field* X *on an orientable manifold of even dimension* m *satisfies* $\delta\mathcal{L}_X\omega = 0$ *for any* m/2 *-form such that* $\delta\omega = 0$ *if and only if* X *is a conformal vector field.*

To determine the invariance group of Maxwell's equation, we use :

a) a theorem of Palais concerning finite dimensional Lie algebras which are generated by complete vector fields;

b) the fact that, in Minkowski space as well as in Segal's model, we know enough explicit Maxwell's fields to determine the Lie algebra of vector fields which stabilize Maxwell's fields;

c) an "ad hoc" reasoning to get all connected components of the group A.

We get :

The invariance group of Maxwell's equations in Minkowski space is Poincaré group extended by dilatations.

One also has a local property: the Lie algebra of C^∞ vector fields on Minkowski space which stabilize Maxwell's fields is the Lie algebra of conformal vector fields which is isomorphic to so(2,4).This algebra contains vector feilds which are not complete. Hence, if one calls conformal group a Lie group with Lie algebra isomorphic to so (2,4), there is no natural global action of a conformal group on Minkowski space. This remark has led us to consider Maxwell's equations in Segal's model. Indeed, it is a compact manifold on which the group SO(2,4) acts naturally. Furthermore, this model can be characterized, up to covering, by the properties of being connected and homogeneous under the action of a conformal group with a certain isotropy.

In this space, we have :

The invariance group of Maxwell's equations in Segal's model is the conformal group of this model.

Furthermore, the space of all C^∞ solutions of Maxwell's equations for free fields in Segal's model can be described in details using a Fourier decomposition of these fields and the computation of all coclosed 1-forms which are eigenforms for the Laplace Beltrami operator on the sphere S^3.

REFERENCES

(1) M. Cahen, S. Gutt : "Invariance des équations de Maxwell", Bull. Soc. Math. de Belgique, Volume en hommage à J. Geheniau (to appear).

(2) M. Cahen, S. Gutt : "Maxwell's equations in Segal's model : solutions and their invariance". Lett. in Math. Phys. 4(1980).

ON THE CLASSIFICATION OF MANIFOLDS FOLIATED BY THE ACTION OF A NILPOTENT LIE GROUP

G. Hector

University of Lille

A homogeneous space of a Lie group G is a manifold admitting a transitive differentiable action of G; these manifolds were the subject of many studies. The actions the orbits of which are the leaves of a foliation form a particularly interesting family of non transitive actions. In order to classify the corresponding manifolds, one is of course obliged to limit oneself to the case of foliations of codimension 1.

Our aim, here, will be to classify the compact manifolds M which admit a *nilfoliation*, i.e. a foliation of codimension 1 of class C^{∞}, transversaly orientable, which is defined by the action ϕ of a nilpotent Lie group G.

Before describing our results, we begin with a short historical account of the subject.

A. The first case to consider is, of course, that of a foliation F on M^{n+1}, defined by a locally free action of $\mathbb{R}^n$. It is not too difficult to see that in this case, F is *nearly without holonomy*, i.e. the holonomy of any non compact leaf is trivial. There are essentially three papers on the subject:

a) In (4), Rosenberg, Roussarie and Weil study the actions of $\mathbb{R}^2$. They show that :

M. Cahen et al. (eds.), Differential Geometry and Mathematical Physics, 31–35.

(i) if F is without holonomy, then M is a circle bundle over the torus T^2. In particular, if the action of $\mathbb{R}^2$ is free (i.e. the leaves of F are planes), then $M = T^3$.

(ii) in general, $M = T^2 \times I$ or M is a torus bundle over S^1.

b) The results for the actions of $\mathbb{R}^n$ are similar (see (5)).

(i) if F is without holonomy, then M is a (n-k)-torus bundle over a manifold B homeomorphic to T^{n-k}.

(ii) if F has compact leaves, then M is a n-torus bundle over I or S^1.

c) Besides, if F is without holonomy (cf. b. (ii)), then M is also homeomorphic to a n-torus or (n-2)-torus bundle over S^1 or T^3 respectively (see (3)).

The results in a) may be obtained as consequences of Stallings' theorem of fibration.

For b) the typical situation is that of a manifold M with boundary with a foliation F tangent to the boundary and with no compact interior leaf. One then constructs a vector field X on M which is tangent to the foliation and such that the limit sets of any trajectory are contained in ∂M. To show that M is a trivial cobordism, it is then enough to modify X so as to make it transverse to F in a neighborhood of ∂M.

It is possible to construct such a field X because all the isotropy groups are equal. This is of course no longer the case for the action of a nilpotent Lie group. Nevertheless, G. Châtelet was able to extend the preceding methods and results to the actions of Heisenberg groups, because for such an action one keeps a good control on the isotropy groups.

However, there is apparently no hope to use the same technique for nilpotent Lie groups in general.

B. In our work, we no longer limit ourselves to locally free actions and we study the actions ϕ of any nilpotent Lie group G, as long as they define a foliation F of codimension 1 (tangent to the boundary).

There are roughly speaking three steps:

a) The fact that any nilpotent Lie group has polynomial growth implies that the same holds for the leaves of a nilfoliation F on a compact manifold M. From this one deduces that F is nearly without holonomy. This allows to limit the study to that of the *models* (cf. (6)), the most important models being those of type 1 (these are nilfoliations (M,F) such that the foliation $\overset{\circ}{F}$ induced by F in the interior $\overset{\circ}{M}$ of M is without holonomy).

b) We then prove a theorem of fibration for the models of type 1. To do this, we use Malcev's theory of homogeneous spaces of nilpotent Lie groups (cf. (4)) and we partially borrow a proof from Châtelet (cf. (7)).

For $y \in M$, one has a natural projection $q_y : G_y \to \pi_1(F_y, y)$ of the isotropy group G_y of G at y onto the fundamental group of the leaf through y, the kernel of q_y being equal to the identity component G_y° of G. Then, Γ_y being the kernel of the holonomy representation of the leaf F_y, we denote by $\hat{\Gamma}'_y$ the smallest connected closed subgroup containing $\Gamma'_y = q_y^{-1}(\Gamma_y)$. We show that the dimension of the Lie algebra $\mathcal{G}_y$ of $\hat{\Gamma}'_y$ is independent of y, equal to k and that the natural inclusion of $\mathcal{G}_y$ into the Lie algebra $\mathcal{G}$ of G defines a continuous map

$$\chi : M \to \mathcal{M}^k(\mathcal{G})$$

where $\mathcal{M}^k(\mathcal{G})$ is the Grassmann manifold of k-planes of $\mathcal{G}$.

Using χ and the action ϕ, we construct a totally integrable subbundle E of T(M) which turns out to be the tangent bundle to the fibers of a locally trivial fibration:

$$\Lambda \to M \overset{\pi}{\to} B \qquad (\star)$$

which we call the *canonical fibration of the model* (M,F). Besides we show that the fiber Λ is a compact nilmanifold and that the base B is endowed with a Reeb foliation (i.e. a foliation the leaves of which are planes or tori) such that the initial foliation F on M is simply the inverse image of R under π.

c) The last step is maybe the most important one : we study the models of type (1.2) which are the models of type 1 such that every leaf of $\overset{\circ}{F}$ is dense in M.

It is not very difficult to show that the fundamental group of M is strictly polycyclic and that M is an Eilenberg-McLane space. Let L be a connected component of ∂M and j the natural inclusion of L into M. We successively show that :

(i) j is a homotopy equivalence and ∂M has two connected components L and L';

(ii) (L,M,L') is a s-cobordism;

(iii) (L,M,L') is a trivial cobordism.

The methods we use are of a topological nature. To prove (i), we use the fact (which seems to be of interest by itself) that any homogeneous space of a polycyclic group has at most two ends. To calculate the Whitehead torsion $\tau(M,L)$, we construct a new fibration of M, this time over S^1, and we apply a result due to D.R. Anderson (cf. (2)).

At last, (iii) is evidently obtained by applying the s-cobordism theorem in large dimensions ; in small dimension we use the fibration (*).

The results can be summarized in the following way:

Theorem. *Let* (M,F) *be a nilfoliation. Then* F *is nearly without holonomy and furthermore,*

(i) if F *is without holonomy,* M *is a locally trivial fibre bundle with base a differentiable manifold homeomorphic to a torus and fibre a nilmanifold.*

(ii) if F *has a compact leaf* L, *then* M *is a fibre bundle over* S^1 *with fibre* L.

In any case, M has the homotopy type of an "infra-solvmanifold". (A manifold is an infra-solvmanifold if it admits as finite cyclic covering a homogeneous space of a solvable Lie group.).

One might of course be tempted to extend the preceding results to the foliations defined by the action of a solvable Lie group.

Unfortunately, such foliations are no longer nearly without holonomy and even if one imposed this extra condition, it is not at all clear that one could construct the analogue of the canonical fibration (★). Indeed, this is constructed thanks to Malcev's results, and it is essential for our purpose.

REFERENCES

(1) H. Rosenberg, R. Roussarie, D.Weil, A classification of closed orientable 3-manifolds of rank two. Ann. of Math., 91 (1970), p. 449-464.

(2) D.R. Anderson, A note on the Whitehead torsion of a bundle modulo a subbundle. Proc. Amer. Math. Soc., 32-2 (1972), 593-595.

(3) D. Tisscher, Locally free actions of $\mathbb{R}^{n-1}$ on M^n without compact orbits. Topology 13 (1974), p. 215-217.

(4) A.I. Malcev, On a class of homogeneous spaces, Amer. Math. Soc. Transl. , 39(1951), p. 276-307.

(5) G. Châtelet, H. Rosenberg, Manifolds which admit $\mathbb{R}^n$ actions. Publ. Math. IHES, 43 (1974), 245-260.

(7) G. Châtelet, Sur les feuilletages induits par l'action de groupes de Lie nilpotents. Ann. Inst. Fourier, 27-2 (1977), 161-190.

HEAT, COLD AND GEOMETRY

P. Iglesias
(CNRS., Marseille)

J.M. Souriau
(Université de Provence, and CNRS. Marseille).

0. INTRODUCTION

Classical and relativistic mechanics can be formulated in terms of symplectic geometry; this formulation leads to a rigorous statement of the principles of statistical mechanics and of thermodynamics.

This analogy also brings to light however certain fundamental difficulties which remain hidden in the traditional approach through some ambiguities.

The "first principle" of thermodynamics can be formulated so as to avoid this ambiguity provided one accepts a detour through the principle of general relativity and the Einstein equations for gravitation.

The mathematical tools used are the theory of symplectic moments, certain cohomological formulae and the concept of distribution-tensor.

As the "second principle" we shall merely show how it is possible, by accepting a particular geometry status for temperature and entropy, to construct a relativistic model of a dissipative continuous medium. This model has the following properties :

a. it is predictive;

M. Cahen et al. (eds.), Differential Geometry and Mathematical Physics, 37–68.

b. all fits solutions satisfy both principles of thermodynamics and admit a detailed balance (energy-impulse, momentum).

c. it contains in particular all equilibrium situations of statistical mechanics, and also the relativistic theory of elasticity.

d. Finally its non-relativistic limit allows one to identify the usual thermodynamic variables and in particular it contains the theory of elasticity, the mechanics of perfect fluids, the theory of heat conduction (Fourier) and the theory of viscosity (Navier).

Nevertheless it is a schematic model which does not take into account phenomena such as capillarity, plasticity, electromagnetic effects, etc.

I. SYMPLECTIC FORMULATION OF DYNAMICS

Consider first of all an elementary dynamical system : a newtonian point mass of mass m, position $\vec{r}$, velocity $\vec{v}$, in a force field $(\vec{r},t) \to \vec{F}$[1]; the triplet $y = (\vec{v},\vec{r},t)$ makes up an initial condition for a *motion* x; y travels through a manifold v, (évolution space); if one puts

$$\sigma_V(dy)(\delta y) = \langle m\, d\vec{v} - \vec{F}\, dt, \delta\vec{r} - \vec{v}\,\delta t\rangle - \langle m\,\delta\vec{v} - \vec{F}\delta t, d\vec{r} - \vec{v}\, dt\rangle \qquad (1.1.)$$

d and δ being two arbitrary variations, the brackets <,> representing scalar product in $\mathbb{R}^3$, one defines on V, a 2-form σ_V of rank 6; the equations of motion become $dy \in \ker(\sigma_V)$[2] ; if $\vec{F}$ is the gradient of some potential, σ_V is a *closed* form (its exterior derivative vanishes); σ_V is thus an absolute invariant integral of the equations of motion, discovered by E. Cartan, but in fact already described explicitly by Lagrange.

The set U of all possible motions has a structure of symplectic manifold (of dimension 6), provided with the closed and reversible 2-form σ_V, whose reciprocal image by the submersion $y \to x$ coincides with σ_V (fig. 1).

Such a scheme can be extended to general dynamical systems

(systems of many particles, spin particles, relativistic mechanics, etc.; see (20)) : in all cases, the space U of motions is a symplectic manifold (thus of even dimension) onto which the evolution space V is projected, each section t = constant of V is a "phase space",; however the identification of phase spaces corresponding to different times is an arbitrary operation, which depends on the system of reference chosen, and therefore is best avoided.

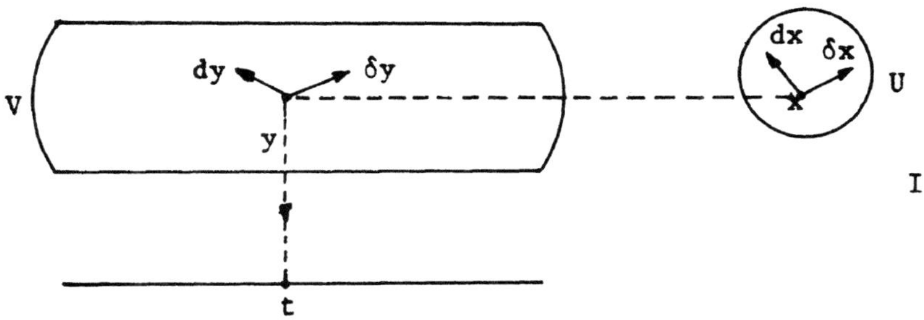

I

II. SYMPLECTIC FORMULATION OF STATISTICAL MECHANICS

In this representation a *statistical state* μ is simply a probability law defined on U (i.e. a positive measure of weight 1); the set Prob(U) of these probability laws is a *convex* set, whose *extremal points* are the classical motions x (identified with the corresponding Dirac measures) (Fig. V).

The completely continuous states are characterized by a *density* of U, which is the product of the *Liouville density* [(3)] by a scalar which can be identified with the classical distribution function[(4)] . The *entropy* of a statistical state μ is defined to be the average value S of -Log ρ for this state; one can define a good class of states, the "Boltzmann states" (23), which make up a convex subset of Prob (U) and for which the integral of -log ρ is convergent; μ → S is a concave function on this convex.

III. THE PRINCIPLES OF THERMODYNAMICS

Statistical mechanics, as it has just been described, is capable of describing various real phenomena, but not *dissipative* phenomena (friction, heat conduction, viscosity, etc.) which make up the study of thermodynamics. The two "principles" of thermodynamics apply in fact only to idealized situations : dissipative transitions, in which a system is in a statistical state μ_{in} *before* the dissipative phenomena and reaches a new statistical state μ_{out} *after* the phenomena. The second principle (Carnot-Clausius) then reads

$$S(\mu_{out}) \geqslant S(\mu_{in}) \tag{3.1}$$

whereas the first principle expresses the *conservation* of *the mean value of the energy* E, which can be written

$$\mu_{out}(x \leftrightarrow E) = \mu_{in}(x \leftrightarrow E) \tag{3.2}$$

taking the measures to be linear functionals.

Both μ_{in} and μ_{out} belong to the convex of Boltzmann states giving a given mean value Q to the energy. It can happen that, on this convex, the concave function S be bounded; let S_Q be then its upper bound. Obviously

$$S(\mu_{in}) \leqslant S(\mu_{out}) \leqslant S_Q \; . \tag{3.3}$$

This gives a majorant of $S_Q - S(\mu_{in})$, known as a function of μ_{in} only, to the *entropy production* $S(\mu_{out}) - S(\mu_{in})$. It can also happen that the maximum of S on this convex be reached at a unique point μ_Q, known as a *Gibbs state* ; if $\mu_{in} = \mu_Q$, the entropy production vanishes and $\mu_{out} = \mu_{in}$: so Gibbs states cannot undergo dissipative phenomena; they constitute what is known as *thermodynamic equilibria.*

IV. COVARIANT FORMULATION OF THE FIRST PRINCIPLE

The foregoing analysis applies to conservative systems; the function $x \leftrightarrow E$, defined on a symplectic manifold U, permits by the Hamiltonian formalism to define a one-parameter group of symplectomorphisms of U[5]; calculations show that this group is lifted to the evolution space V by the group of time translations; in the case of a single particle

$$\vec{v} \to \vec{v}\ ,\ \vec{r} \to \vec{r}\ ,\ t \to t + Cte \qquad (4.1)$$

this is usually expressed somewhat incorrectly by saying that "time and energy are conjugate variables"[6].

Clearly translations (4.1) are linked to a particular frame: the first principle, as stated, does not respect relativistic covariance, even galilean[7]; there must therefore exist a statement avoiding this drawback.

A radical solution is to replace the group (4.1) by the complete galilean group[8], or else in the relativistic case, by the Poincaré group[9].

The action of these groups on U by symplectomorphisms is defined in a natural way if the dynamical system is isolated; otherwise one considers a partial system, to which the "mechanism" made up by the given exterior system leaves only the symmetry corresponding to a subgroup of the galilean (or Poincaré) group. For example a fixed box containing a gas, which restricts the gas to the subgroup (4.1); but also a centrifugal machine, etc.

Let G then be the group of symmetries; we seek a quantity which plays the same role with respect to G as does the energy with respect to the group (4.1).

It is sufficient to achieve this to consider *all one-parameter subgroups of* G; each one will be characterized by an element Z of the Lie algebra $\mathcal{G}$ of G; to each one will correspond a hamiltonian which will be denoted by M(Z). Inspection shows that one can choose the additive constant which appears in each hamiltonian in such a way that the correspondence $Z \mapsto M(Z)$ be *linear;* M becomes thus a linear form on $\mathcal{G}$, thus an element of its dual $\mathcal{G}^\star$; there exists therefore an application $x \mapsto M$ of U into $\mathcal{G}^\star$; the variable M will be called the *moment* of the group; naturally therefore one replace the first principle (3.2) by the statement (19)

$$\mu_{out}(x \mapsto M) = \mu_{in}(x \mapsto M) \qquad (4.2)$$

without changing the second principle (3.1); the conclusions are similar. On the convex of Boltzmann states satisfying $\mu(x \mapsto M)=Q$[10], there may exist a "Gibbs state" μ_Q having the largest entropy S_Q; as before, one obtains a majorant to the entropy production in a dissipative transition; and one arrives at the conclusion that Gibbs states are no longer susceptible to dissipative phenomena.

The distribution function of these Gibbs states is the exponential of an affine function of M, which can be written

$$\rho = e^{M\theta - z} \tag{4.3}$$

θ being an element of G ("geometric temperature"), z a number ("Planck's thermodynamic potential", see (VI)) which is obtained in terms of θ by writing that the weight of μ is 1,

$$z = \log \int_U e^{M\theta} \lambda(x)\, dx \,, \tag{4.4}$$

λ being the Liouville measure; z is a *convex* function of θ, which turns out to be the *Legendre transform* of $Q \mapsto -S_Q$:

$$dQ\theta = - dS \,,\ Q\, d\theta = dz \,,\ Q\theta = z - S \,,\ \forall\ d. \tag{4.5}$$

All the classical formulae of thermodynamics are thus generalised but now the variables are provided with a geometrical status. For instance, the geometrical temperature θ, an element of the Lie algebra of the Galilea or Poincaré groups, can be interpreted as the *field of space-time vectors;* in the relativistic version, $\theta(x)$ is a *time like vector*, its orientation characterises the "arrow of time"; its direction is the 4-velocity of the equilibrium referential; its (Minkowski) length is $\beta = \frac{1}{kT}$ (k : Boltzmann's constant, T : absolute temperature). This temperature-vector had already been suggested by Planck in order to study relativistic thermodynamics, but its galilean counterpart is quite as relevant.

The formulae thus obtained can be applied correctly to a number of real situations : equilibrium of spin particles, centrifugal machines, rotation of celestial bodies, etc.

Furthermore new relations appear, linked to the non-commutativity of the group G, which give rise to some predictions; thus, under weak hypotheses, one can predict the existence of a *critical temperature* for an isolated system, beyond which no equilibrium state will exist; this fact is probably important in astrophysics (supernovae). For further details see (20) and (23).

V. GRAVITATIONAL SUSCEPTIBILITY

The covariant formulation (4.2) of the first principle removes thus a paradox, and at the same line increases the practical value of thermodynamics. However it leaves a conceptual problem.

During a dissipative transition, statistical mechanics is necessarily violated since $\mu_{out} \neq \mu_{in}$; the dynamical variable energy (or more generally the moment) *changes spectrum during the transition*[11]. As one can no longer appeal to conservation laws of classical or statistical mechanics, it is necessary to bring in *other laws of nature* in order to understand how the mean value of the energy is *conserved*, or more simply how it is *memorized*.

Somewhat unexpectly the answer is provided by general relativity; we shall see how in §7, after a study of preliminary concepts. Consider a dynamical system evolving in a *gravitational field*, field which is characterized in general relativity by its *potentials* $g_{\mu\nu}$; the space of all motions is always a symplectic manifold U, whose structure depends on the field.

Now we choose a compact K of space-time E_4 (see fig. II) wherein we perturb the $g_{\mu\nu}$. The new space of motions U' is still a symplectic manifold, which can be connected to U by the technique of *diffusion;* this technique will be described in the case of a spinless particle, whose motion is characterized by the world line; if this line does not meet K, it characterizes a motion equally in U as in U'.

Consider now a motion in U, which we shall denote by X_{in} whose world line centers at some time into K; with the perturbed potentials, the line will deviate from the initial motion (dotted line in figure). When it leaves K, it takes a new path which can be identified to an element X_{out} of U; the correspondence $X_{in} \rightarrow X_{out}$, which characterizes globally *diffusion by scattering* is a local symplectomorphism of U (because U and U' are each symplectic, and their structures can be obtained by starting from the same evolution space). Such a quantity can be characterized by a certain dynamical variable called the diffusion *eikonal*.

We are interested here in *infinitesimal diffusion :* if one gives to the $g_{\mu\nu}$ a variation $\delta g_{\mu\nu}$ which vanishes outside K, the initial motion will undergo a displacement $\delta x = F(X)$ which derives from a certain hamiltonian ϕ (see §4); ϕ, so defined to within an additive constant, can be completely determined by putting it equal to zero on all paths which do not cross K[12].

For any motion $x \in U$, let T_x be the application which establishes a correspondence between ϕ and the tensor field $x \nrightarrow \delta g$ $(x \in E_4)$, T_x is a linear application, thus a priori a distribution[13]; knowing T_x allows one to predict how the particle will react to any "slight" modification of the gravitational field: this is why T_x will be called the *gravitational susceptibility* of the particle in the motion x.

We now use the *general relativity principle :* it states that a diffeomorphism A of time-space E_4, acting simultaneously on the potential (according to the standard formulae of differential geometry) and in the motion (here by direct image of the world line) is *unobservable* (see (21)). Let us choose A so that it leaves unchanged the points outside the compact K (fig. III); it modifies the potentials only in K, and its action in the particle leaves unchanged those parts of the world line outside K, the corresponding diffusion by scattering is thus zero.

Let us apply this result to the case $A = \exp(sF)$ (see 2.4), $s \in \mathbb{R}$, F = vector field vanishing outside K. One sees that $T_x(x \nrightarrow \delta g)$ vanishes if g is the derivative with respect to s, at s = 0, of the reciprocal image of A by $x \nrightarrow \delta g$; this variation by definition is the *Lie derivative* of g associated to the vector field $\delta X = F(X)$; we shall denote it as $\delta_L g$. Generally speaking a

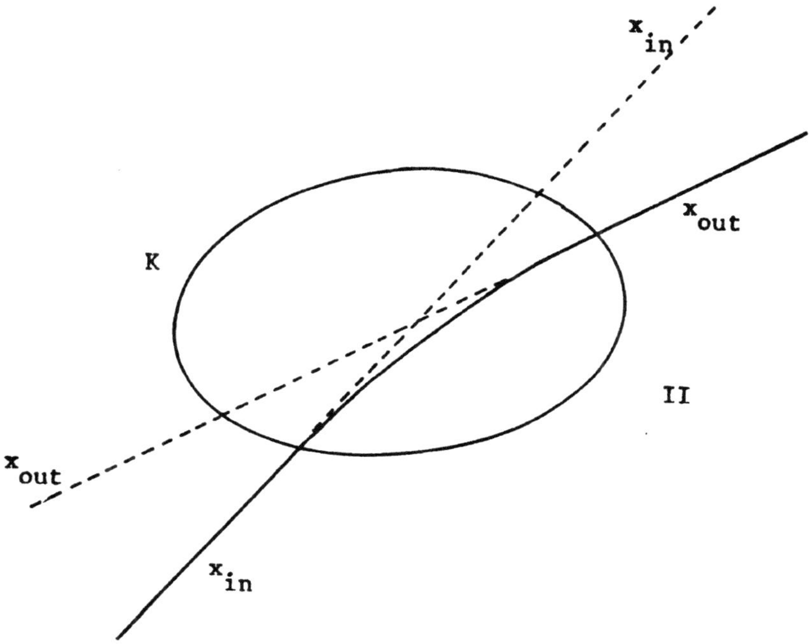
x_{in}
x_{out}
K
II
x_{out}
x_{in}

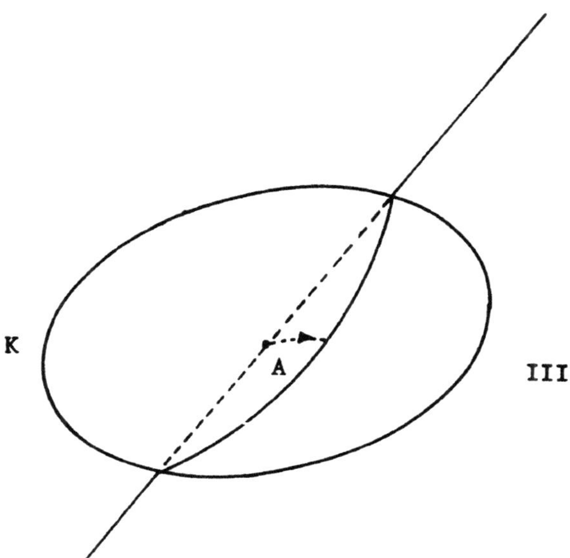
K
A
III

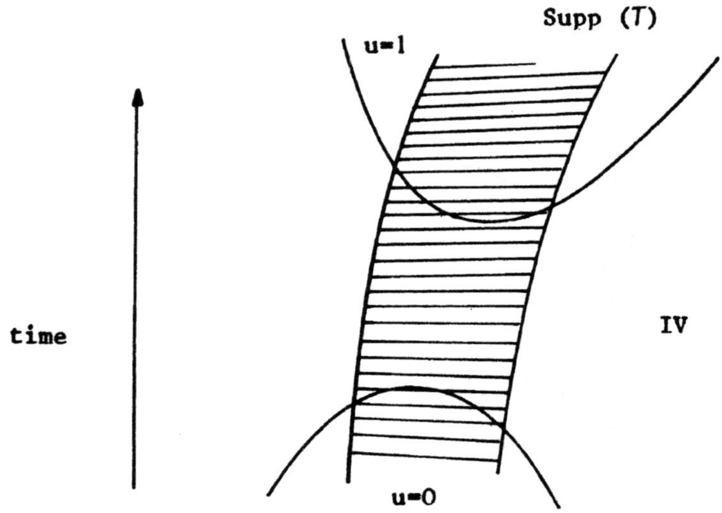

Eulerian distributions

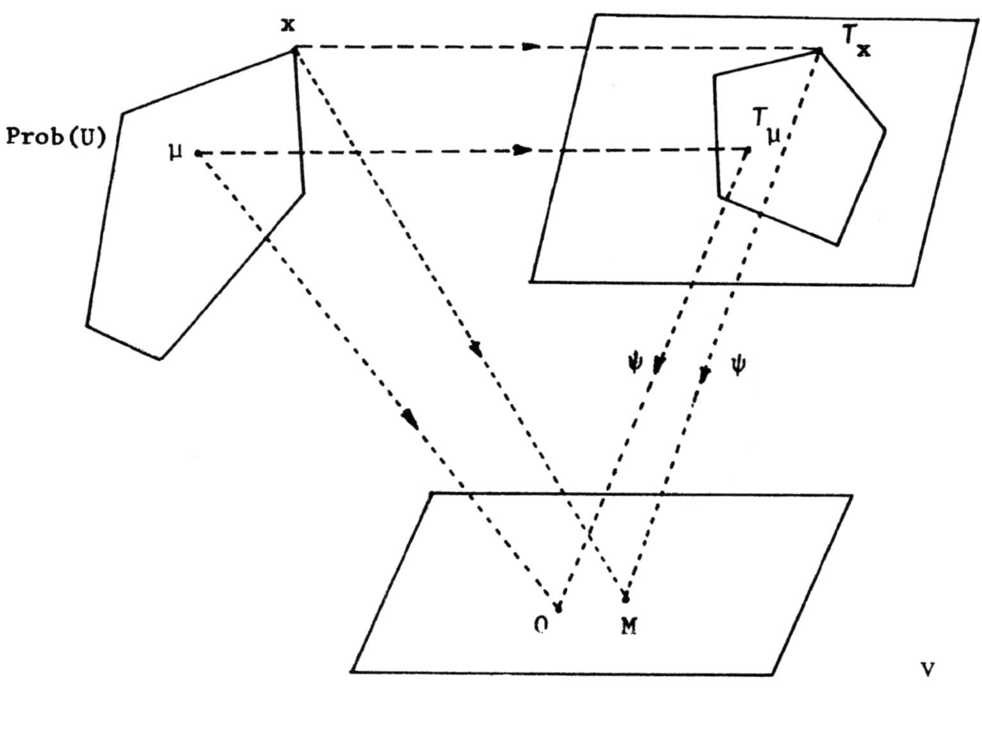

distribution T will be said to be *eulerian* if it satisfies the condition

$$T(X \leftrightarrow \delta_L g) = 0 \text{ for any field } X \leftrightarrow \delta X \text{ under compact support;} \tag{5.1}$$

we know therefore that $x \leftrightarrow T_x$ is an application of the space of motions U into the vector space of eulerian distributions of E_4 (fig. V). Under certain hypotheses, a eulerian distribution allows one to associate a *conserved quantity* to any Killing vector Z of the metric g[13]; we shall give a brief description of this procedure in the case where E_4 is Minkowski space, and consequently Z is an element of the Lie algebra of the Poincaré group; the associated quantity is

$$I = T(X \leftrightarrow \delta_L g) \text{ where } \delta X = uZ(x) \ , \tag{5.2}$$

U being a function taken to be equal to zero in the past and equal to one in the future (fig. IV).

Contrary to what one may think by studying (5.1) and (5.2), I is not necessarily vanishing, because $X \leftrightarrow u\, Z(x)$ is not a field with compact support; but the eulerian condition (5.1) allows one to show that I *does not depend on the choice of* u, by making some assumptions on the behaviour of T at infinity[14]. One can thus calculate I by making u jump from 0 to 1 in a small neighborhood of a space-like surface; the fact that the result is independent of the choice of this surface expresses the "preserved quantity" character of I.

Clearly the application $Z \leftrightarrow I$ so defined is linear; it associates to T an element $\Psi(T)$ of the dual $G^\star$ of the Lie algebra G of the Poincaré group. It is obvious that the application

$$x \leftrightarrow M = \Psi(T_x) \tag{5.3}$$

is a moment of the Poincaré group (§4); we shall now find another property by *equivariance* considerations.

It is clear that the Poincaré group G acts on tensor fields with compact support, thus on distribution-tensors according to the formula

$$a(T)(a(X \leftrightarrow \delta g)) = T(X \leftrightarrow \delta g) \ , \quad \forall\, a \in G. \tag{5.4}$$

It also acts on vector fields with compact support, and the Lie derivative of g is equivariant for this action, so that G acts on eulerian distributions. Finally G acts on Killing vectors, and this action coincides with the *adjoint representation* of G on $\mathcal{G}$.

Let us now suppose that a be an element of the *orthochrone* subgroup $G^{\uparrow}$ (16): the foregoing constructions show that

$$\Psi(a(\mathcal{T}))(a(Z)) = \Psi(\mathcal{T})(Z) \tag{5.5}$$

and $\forall\, x \in U$, that

$$a(\mathcal{T}_x) = \mathcal{T}_{a(x)} \tag{5.6}$$

whence

$$\Psi(\mathcal{T}_{a(x)})(a(Z) = \Psi(\mathcal{T}_x)(Z). \tag{5.7}$$

This last formula has a *cohomological* interpretation : it expresses the vanishing of a certain symplectic cocycle and entails that (see(19)).

$$M[Z,Z'] = \sigma(Z(x))(Z'(x)) \quad , \forall\, Z,Z' \in \mathcal{G}. \tag{5.8}$$

This result allows one to fix the arbitrary constant which appeared in M (because the Lie algebra of the Poincaré group is equal to its derived algebra); it shows in particular that if Z is an infinitesimal time translation, then the integral I (5.2) is equal to the *relativistic energy* $E = mc^2$ of the system in the motion in question.

We have thus *factorized* the "moment" application $x \mapsto M$ by the composition of $x \mapsto T_x$ and $T_x \mapsto M$ (fig. V); this is the result which will be essential for thermodynamics. Let us indicate in detail what these results become for a single particle; the symplectic form (1.1), in general relativity, reads

$$\sigma_V(dy)(\delta y) = m\, g_{\mu\nu}[dX^\mu\, \hat{\delta} U^\nu - \delta X^\mu\, \hat{d} U^\nu] . \tag{5.9}$$

In this case an initial condition y is a pair (X,U), X belonging to the world line, U being the unit tangent vector; the carets ^ represent covariant differentiation.

The calculation (somewhat technical) of the gravitational susceptibility yields

$$T_x(X \leftrightarrow \delta g) = \int \frac{1}{2} m \quad \delta g_{\mu\nu} \; p^{\mu} \frac{dX^{\nu}}{ds} \, ds \tag{5.10}$$

$s \leftrightarrow X$ being an arbitrary parametrization of the world line of the particle (from the past towards future). We see that the distribution T_x is a *measure*, having as support this world line. It so happens that we know *all* the eulerian measures supported by a curve; they can be written as

$$T(X \leftrightarrow \delta g) = \int \frac{1}{2} \delta g_{\mu\nu} \; p^{\mu} \frac{dX^{\nu}}{ds} \, ds \tag{5.11}$$

with the supplementary conditions

$$\frac{dX}{ds} \text{ parallel to } P \; ; \; \frac{\hat{dP}}{ds} = 0 \tag{5.12}$$

(the proof can be found in (2.1)); these conditions imply that the curve be a geodesic : a well-known fact for particles, which can also be found by using d'Alembert's principle $dy \in \ker(\sigma_V)$ in the form (5.9). The 4-momentum $P = mU$ appears thus as an element of the gravitational susceptibility; in the case of Minkowski space, the preserved quantity associated with an element Z of the Lie algebra G of the Poincaré group is

$$I = g_{\mu\nu} \; p^{\mu} \; Z(X)^{\nu} \tag{5.13}$$

X being chosen arbitrarily on the world line; by varying Z in G , one can display energy, momentum, orbital momentum, etc.

The structures we have just displayed in the simplest case can be extended to a great variety of circumstances.

They can be transposed to the *classical mechanics* case; to each motion x we still associate a distribution T_x; the eulerian condition is expressed, no longer by a Lie derivative, but by a certain *connection* which takes into account the gravitational field in its newtonian form. One notes that the conserved quantities associated with the null field case bring out a Lie algebra of dimension 10 which is *not* that of the Galilean group, but somewhat paradoxically that of the *Caroll group*, which is a contraction of the Poincaré group obtained by letting the speed of light c go to zero. This phenomenon can be put together with the impossibility of choosing the galilean group moments in such a way as to satisfy formula (5.8) : an obstruction appears which is a class of symplectic cohomology and which is measured by the total mass of the system, thus non-vanishing.

Physically speaking, these facts indicate that in the formulation of classical mechanics by eulerian distribution, *mass conceals energy.*

The same method can be used to treat spin particles, both in classical and relativistic mechanics. The gravitational susceptibility involves, together with the 4-momentum P^{μ} , the antisymmetric *spin tensor* $s^{\mu\nu}$.

This method allows one to obtain in a simple way the collision and desintegration rules for particles: one has simply to write that the sum of the gravitational susceptibilities carried by the various world lines is still a eulerian distribution. This method can be extended to *electrodynamics* : one calculates the gravitational susceptibility for a simultaneous variation of the gravitation potentials $g_{\mu\nu}$ and the *electromagnetic potentials* A_{ρ}; for particles this susceptibility introduces, together with the 4-momentum and the spin tensor, the electric charge and the magnetic moment. The general relativity principle, which affected the group of diffeomorphisms in space-time, is generalized to an electromagnetic group; consequently the eulerian condition becomes

$$\mathcal{T}(X \leftrightarrow (\delta g, \delta A)) = 0 \text{ if } \quad \delta g = \delta_L g \ , \ \delta A = \delta_L A + \frac{\delta \alpha}{\delta X} \qquad (5.14)$$

$X \leftrightarrow \delta X$ and $X \leftrightarrow \alpha$ being a vector field and a scalar field with compact support respectively.

Note that these structures become particularly simple when written in the 5-dimensional space-time of Kaloza.

VI. LOCALISATION OF STATISTICAL STATES

Let μ be a statistical state of a dynamical system, i.e. a probability law of the manifold U of motions (fig. V).

If $X \nrightarrow \delta g$ is a variation with compact support of the gravitation potentials, we can establish a correspondence between every motion $X \in U$, and a diffusion Hamiltonian $\phi = T_x(X \nrightarrow \delta g)$; ϕ is a function of x, i.e. a dynamical variable; the state μ will be called *localisable* if this function is μ-integrable $X \nrightarrow \delta g$; we shall then put

$$T_\mu(X \nrightarrow \delta g) = \int_U T_x(X \nrightarrow \delta g)\mu(x)dx \quad ; \tag{6.1}$$

this quantity is the *mean value* of ϕ is state μ.

One can check immediately that :

- the set of all localisable statistical states is a sub-convex of Prob(U), and contains its extremal points. (6.2)

-If μ is localisable, T_μ is a *eulerian distribution.* (6.3)

- In the case of Minkowski space, the element $\Psi(T_\mu)$ of g (see §5) is equal to the mean value Q of M in state μ(see fig. V). (6.4)

- It would appear that localisable states are the only ones met in nature; in particular Gibbs states are localisable. If the state μ has a distribution function of class $\mathbb{C}^\infty$ (on U), T_μ will be a *completely continuous* distribution (on E_4), which will be written

$$T(X \nrightarrow \delta g) = \int_{E_4} \frac{1}{2}\ \delta g_{\nu\rho}\ T^{\nu\rho}\ (X)\ dX \tag{6.5}$$

The $T^{\nu\rho}$ being densities ($T^{\nu\rho} = T^{\rho\nu}$); these $T^{\nu\rho}$ are the components of a *tensorial density* in the Brillouin sense; they can also be written as $T^{\nu\rho}\lambda$, $T^{\nu\rho}$ now being the components of a symmetric tensor, and λ the riemannian density of space-line ; (6.5) becomes

$$T_\mu(X \leftrightarrow \delta g) = \int_{E_4} \frac{1}{2} T^{\nu\rho} \, \delta g_{\nu\rho} \, \lambda(X) \, dX \tag{6.6}$$

the eulerian condition (5.1) is obtained from the Killing formula

$$[\delta_L g]_{\alpha\beta} = g_{\alpha\gamma} \, \hat{\partial}_\beta \delta X^\gamma + g_{\gamma\beta} \, \hat{\partial}_\alpha \, \delta X^\gamma \tag{6.7}$$

one easily finds

$$\hat{\partial}_\nu T^{\nu\rho} = 0 \tag{6.8}$$

where one recognizes the relativistic form of the Euler equations proposed by Einstein (4).

Consequently the localisation of a statistical state allows one to interpret it with the assistance of a continuous media, whose $T^{\nu\rho}$ defined by t6.6) make up the *energy tensor* and are *automatically* solutions of the Euler equations. This interpretation can be confirmed by detailed calculations; thus in the case of a particle, the T^{oo} component, which is interpreted as the *specific mass*, is the mean value of the relativistic mass $m/\sqrt{1-V^2/C^2}$ in a volume element in the neighborhood of the point X considered; the pressure, or more generally, the constraint tensor is interpreted as a measure of the random character of the speeds of the motions going through X; etc.

Let us consider the case of a system of N relativistic spinless particles of mass m, making up a Gibbs equilibrium in a box of volume V at a temperature T. In a frame linked to the box, the $T^{\nu\rho}$ tensor is diagonal, and can be expressed in terms of a *specific mass* ρ and a pressure P given by

$$\rho = \frac{Nm}{V} \frac{G''(x)}{-G'(x)} \quad P = \frac{Nm}{V} \frac{G''(x) - G(x)}{-3G'(x)} \tag{6.9}$$

where $x = m/kT$, and G is defined by

$$G(x) = \int_1^\infty e^{-xs} \sqrt{s^2-1}\, ds. \tag{6.10}$$

It so happens that $G(x) = \frac{K_1(x)}{x}$, K_1 being the modified Bessel function of order 1; G satisfies therefore the differential equation

$$G''(x) + \frac{3}{x} G'(x) - G(x) = 0 \tag{6.11}$$

whence

$$pV = NkT; \tag{6.12}$$

one recovers thus *exactly* the classical perfect gas law (Boyle-Mariotte-Charles-Gay-Lussac-Avogadro - Boltzmann); the first terms of the asymptotic expansion of K_1 (27) yield the formula

$$Q = \rho V = Nm + \frac{3}{2} NkT = \frac{Nm}{-2G'(m/kT)} \int_1^\infty e^{-ms/kT} (s-1)^{\frac{5}{2}} (s+1)^{\frac{1}{2}} ds. \tag{6.13}$$

Here we have set c=1, the first term is the mass at absolute zero; the second term is the *mean classical value of the energy*, which allows one to calculate the specific heat of a monoatomic gas; the third (positive) term is the relativistic correction.

One also obtains Planck's thermodynamic potential (4.4)

$$z = N \log(-4\pi V m^3 G'(m/kT) F \tag{6.14}$$

which indeed is a convex function of $\beta = 1/kT$; formula (4.5) then yields the entropy(17)

$$S = Kz + \frac{Q}{T} \tag{6.15}$$

It is remarkable that the $T^{\nu\rho}$ tensor, which has been constructed as characteristic of the *gravitational susceptibility*, also characterizes the *gravitational action* of matter defined by the statistical state μ; for it is indeed this tensor which appears in the right- hand side of the Einstein gravitation equations :

$$R^{\nu\rho} - \frac{1}{2} Rg^{\nu\rho} + \Lambda\, g^{\nu\rho} = 8\pi G T^{\nu\rho} \tag{6.16}$$

(Λ cosmological constant; G = Newton's constant; c = 1).

Using the definition (6.1), these equations can be written

$$\int_{E_4} \delta Z(X)dX = \int_U T_x(X \leftrightarrow \delta g)\ \mu(x)dX\ ,\ \forall\ \delta g \qquad (6.17)$$

Z being the lagrangian density of the gravitational field

$$Z = \frac{2\Lambda - g^{\nu\rho}R_{\nu\rho}}{8\pi G}\ \lambda(\lambda = \text{riemannian density}) \qquad (6.18)$$

in this form, one notes that the distribution defined by the first member is *automatically* eulerian.

All this approach can be extended to the electromagnetic case; formulae (6.6) and (6.8) become

$$T_\mu(X \leftrightarrow (\delta g, \delta A)) = \int_{E_4} [\tfrac{1}{2}\ T^{\nu\rho}\delta g_{\nu\rho} + J^\sigma \delta A_\sigma]\ \lambda(X)dX \qquad (6.19)$$

and

$$\hat{\partial}_\nu T^{\nu\rho} + F_\nu{}^\rho J^\nu = 0 \qquad \hat{\partial}_\sigma J^\sigma = 0 \qquad (6.20)$$

respectively.

Einstein's equations (6.16) are replaced by the coupled Einstein-Maxwell equations; the 4-vector J^σ is interpreted as the current-charge density.

Applying the preceding formulae to statistical states of particles with spin having a magnetic moment allows one to recover the principal characteristics of ferromagnetism (magnetic equivalence magnet-solenoid); gyromagnetic effects; magnetostriction ...) (see (21) and (23)).

VII. GRAVITATIONAL INTERPRETATION OF THE FIRST PRINCIPLE

Consider the case of a dissipative transition $\mu_{in} \to \mu_{out}$, and let us suppose there exists a *eulerian distribution* T that coïncides with $T\mu_{in}$ before the *dissipative phenomena* and with $T\mu_{out}$ afterwards; in other words a distribution T that interpolates between $T\mu_{in}$ and $T\mu_{out}$.

We can then associate to Q a conserved quantity Q which will acquire the same value for $T\mu_{in}$ as for $T\mu_{out}$, as it can be calculated at an arbitrary time; we also know that for $T\mu_{in}$, Q is equal to the mean value of the Poincaré moment M; similarly for $T\mu_{out}$; consequently the *first principle*, in its covariant form (4.2) will be assured. We only need to set (notation 6.17)

$$T(X \leftrightarrow \delta g) = \int_{E_4} \delta\mathcal{L}(X)dX \; ; \tag{7.1}$$

indeed it is known that T is a eulerian distribution that interpolates between $T\mu_{in}$ and $T\mu_{out}$, as the Einstein equations (6.16) are valid before and after the dissipative phenomena[18]. It is thus the gravitational potentials g that remember the mean value of the moment M and which guarantee the validity of the first principle (in its covariant form (4.2)).

VIII. THE DISSIPATIVE MEDIUM MODEL

Relativity, thermodynamics, matter and geometry

Consider a Gibbs state of a particle in a symmetrical gravitational field; there exists in space-time a conservative current[19] S whose integral on a space-like hypersurface is equal to the statistical entropy

$$S^{\mu} = T^{\mu\nu}\theta_{\nu} - F\theta^{\mu} \tag{8.1}$$

T being given here by the construction explained in §6, F is the specific free energy of the system [20] and θ is the *temperature vector*. The conservation of the *entropy current* amounts to the equation

$$\partial_{\mu} S^{\mu} = 0. \tag{8.2}$$

In the case of a dissipitative continuous medium, we shall assume as so many authors (2), (3), (8) ... (12) that the geometrization of the second principle is obtained by the permanence, out of equilibrium, of the entropy current, whose flux S satisfies then

$$\hat{\operatorname{div}}\, S = \hat{\partial}_{\mu} S^{\mu} \geqslant 0. \tag{8.3}$$

At any point div S is interpreted as *the specific entropy production*.

Still in the case of dissipative processes, we shall also assume the permanence of the temperature vector, a less strong hypothesis than the often adopted one of local thermodynamic equilibrium.

The duality between the momentum-energy current and the universe metric on the one hand, and the geometrical nature of the entropy current and the temperature vector on the other suggests there exists a duality entropy-temperature that a complete geometrization of the second principle should clarify.

relativity	thermodynamics	
T	S	matter
$\updownarrow$	$\updownarrow$	
g	θ	geometry

The vectorial character of θ makes it the infinitesimal generator of a one parameter group and thus gives it a strictly *kinematic* role. Its lines of current are the molecules, their abstract set makes up a manifold V_3 called the reference body which corresponds to our three dimensional intuition of "space".

Programm for a model

To construct a model we have to choose a system of fundamental variables from which will be written the equations of motion. What we shall ask of a *thermodynamic* model of a continuous medium is that it takes into account the *kinematic variables*

$$(g,\theta)$$

and the dual variables

$$(T,S)$$

in such a way that all the solutions of the equations of motion satisfy the two principles of thermodynamics

$$\hat{\partial}_\mu T^{\mu\nu} = 0 \text{ and } \partial_\mu S^\mu \geqslant 0 \cdot$$

Furthermore we shall require that the motions contain as special case the Gibbs equilibria for which θ is a Killing vector and S is given by (8.1) and $\partial_\mu S^\mu = 0$.

The characterization of equilibria through infinitesimal isometries emphasizes the role of the tensor γ defined as

$$\gamma = \frac{1}{2}\,\delta_L g \;;\; \delta X = \theta$$

i.e.

$$\gamma_{\mu\nu} = \frac{1}{2}[\hat{\partial}_\mu \theta_\nu + \hat{\partial}_\nu \theta_\mu]$$

γ will be called the *friction tensor.*

It seems reasonable to interpret γ as the source of *dissipative phenomena* as thermodynamic equilibria are characterized by $\gamma = 0$.

A simple phenomenological model satisfying this programm

Classical thermodynamic media are characterized by a function of state, the dissipation function (ex. cf (18)) relating the constraints to the amount of deformation. By noticing that the amount of deformation and the constraint are the spatial parts of the friction tensor γ and the momentum-energy tensor T respectively, it seems reasonable to generalize this hypothesis to relativity by assuming the existence of a *generating function* ϕ[(21)] relating γ and T.

More precisely we suppose that

ϕ depends on the variables $(x, g, \theta, \gamma, q, \frac{\partial q}{\partial x})$ where q represents the molecule of V_3 going through x

$$T^{\mu\nu} = \frac{\partial\phi}{\partial\gamma_{\mu\nu}}\,, \text{ i.e. } \delta\phi = T^{\mu\nu}\delta\gamma_{\mu\nu}, \text{ for any variation} \qquad (8.6)$$

of γ, the other variables kept fixed.

In order to satisfy the principle of general relativity, we shall suppose that ϕ is invariant[(22)] under diffeomorphisms of V_4. We take this statement to be the rigorous expression of the *principle of objectivity* or of *material indifference,* proposed by many authors in the framework of classical mechanics (16),(17). Calculations show that ϕ depends only on the following variables:

molecule	q	
reciprocal of the temperature	$\beta = [g_{\mu\nu}\theta^\mu\theta^\nu]^{1/2}$	
conformation	$h^{ij} = g^{\mu\nu}\partial_\mu q^i \partial_\nu q^j$	
deformation rate	$k^{ij} = -2\gamma^{\mu\nu}\partial_\mu q^i \partial_\nu q^j$	(8.7)
thermal field	$X^i = 2\partial_\mu q^i \gamma^\mu{}_\rho \theta^\rho$	
thermal velocity	$a = \frac{1}{\beta}\gamma_{\mu\nu}\theta^\mu\theta^\nu$	

We have proved the following theorem (22), (7) :

if ϕ is convex on γ ;

if there exists a state function F which depends only on q, β, Q, such that

$$\frac{\partial\phi}{\partial\gamma_{\mu\nu}} = \frac{\partial F}{\partial g_{\mu\nu}} \quad (23) \tag{8.8}$$

and if one sets (24)

$$S^\mu = T^{\mu\nu}\,\theta_\nu - F\theta^\mu \tag{8.9}$$

then the equations of motion defined by the Einstein equations

$$R^{\mu\nu} - \frac{1}{2} R g^{\mu\nu} = 8\,\pi\, G T^{\mu\nu} \tag{8.10}$$

generate the realization of the two principles of thermodynamics

$$\hat{\partial}_\mu T^{\mu\nu} = 0 \quad \text{and} \quad \hat{\partial}_\mu S^\mu \geqslant 0.$$

IX. INTERPRETATION OF THE MODEL

The friction tensor γ

The thermodynamic variables k, χ and a of table 8.7 are built up from the friction tensor γ; one can show that, along the motion, they take as values :

$k^{ij} = \frac{dh^{ij}}{ds}$, with $\frac{dx}{ds} = \theta$; k measures thus the variation, along the lines of current, of the deformation which justifies it being called the rate of deformation.

$a = \frac{d\beta}{ds}$; $\frac{dx}{ds} = \theta$; "a" measures the variation of temperature along the lines of current.

χ^i is the relativistic equivalent of the classical temperature gradient.

Momentum-energy tensor and entropy

The calculation of T and S yields

$$T_{\mu\nu} = \alpha U_\mu U_\nu - 2\partial_\mu q^i \partial_\nu q^j \Lambda_{ij} + (U_\mu \partial_\nu q^i c_i + U_\nu \partial_\mu q^i c_i)$$
$$S^\mu = (\alpha - F)\theta^\mu + \beta g^{\mu\nu}\partial_\nu q^i c_i \tag{9.1}$$

where

$\alpha = \dfrac{\partial\phi}{\partial a}$ is the internal energy in the normal sense of thermodynamics.

$c_i = \dfrac{\partial\phi}{\partial\chi_i}$ is the heat convexion.

$\Lambda_{ij} = \dfrac{\partial\phi}{\partial k^{ij}}$ is the constraint.

The expression for the entropy shows that the non convective heat remains, as in usual thermodynamics, the difference between internal and free energy; furthermore the model distinguishes quite naturally that part of the entropy flux vector which is proportional to θ and its orthogonal part, the convective heat flux.

X . NON DISSIPATIVE LIMIT OF THE MODEL

It follows from the strict convexity of the generating function ϕ that the only non dissipative motions of this model are the thermodynamic equilibria for which the friction vanishes. Nevertheless the hypothesis ϕ affine in γ gives the non dissipative approximation for which every motion satisfies

$$\hat{\partial}_\mu \delta^\mu = 0. \tag{10.1}$$

The energy-momentum tensor becomes

$$T_{\mu\nu} = \beta \frac{\partial F}{\partial\beta} U_\mu U_\nu + F g_{\mu\nu} - 2\partial_\mu q^i \partial_\nu q^j \frac{\partial F}{\partial h^{ij}} \tag{10.2}$$

and the entropy

$$S^\mu = \frac{\partial F}{\partial\beta}\theta^\mu \tag{10.3}$$

and one recovers usual thermodynamics.

Introducing a matter current characterized by its flux N

$$N^{\mu} = nU^{\mu}$$
$$\partial_{\mu} N^{\mu} = 0 \qquad (10.4)$$

allows one, by putting

$$W = F - n\, c^2 \qquad (10.5)$$

to write

$$T_{\mu\nu} = (nc^2+Q)U_{\mu}U_{\nu} + wg_{\mu\nu} - 2\partial_{\mu}q^{i}\partial_{\nu}q^{j}\frac{\partial}{\partial h^{ij}} \qquad (10.6)$$

$$S^{\mu} = Q\theta^{\mu}\ ,\ Q = \beta\frac{\partial W}{\partial \beta}$$

The internal energy α becomes then

$$\alpha = nc^2 + Q + W \qquad (10.7)$$

and Q can be written, by setting $S^{\mu} = SU^{\mu}$ and $\beta = \frac{1}{kT}$,

$$Q = \frac{S}{kT}\ .$$

The expression (10.6) for the energy-momentum tensor allows one to interpret W as the elastic energy of the relativistic theory of elasticity. Formula (10.8) leads to the interpretation of Q as the usual heat of thermodynamics, and one recovers in (10,7) the usual expression for the internal energy.

Thus the non dissipative limit of the model encorporates completely the usual definitions and equations of thermodynamics. Furthermore the equations of motion $\partial_{\mu}T^{\mu\nu} = 0$ and $\partial_{\mu}S^{\mu} = 0$ express the relativistic balance of thermodynamics of reversible process of energy, entropy ... and *one recovers the variational theory of elastic media.*

The equations $\hat{\partial}_{\mu}S^{\mu} = 0$ and $S^{\mu} = \beta\frac{\partial F}{\partial \beta}\theta^{\mu}$ allow one to eliminate the temperature β as independent variable (by using a Legendre transformation). It is then possible to construct a Lagrangian density depending only on q and h from which the equations of motions can be derived. This lagrangian coincides with that of the relativistic theory of elasticity (24). Going from this lagrangian to the classical limit gives Hamilton's principle as applied to continuous media.

Perfect fluids and statistical mechanics

The perfect fluid is obtained by letting F depend on h only through its determinant, i.e. the matter density n (10.4) (see (24), (15)). In this case, we have

$$
\begin{aligned}
T &= \frac{\partial[\beta F]}{\partial\beta} U_\mu U_\nu - \left(\frac{\partial F}{\partial n} n - F\right)(g_{\mu\nu} - U_\mu U_\nu) \\
S^\mu &= \sigma N^\mu \ , \quad \sigma = \zeta - \frac{\partial\zeta}{\partial\beta}\beta \ , \zeta = -\frac{\beta F}{n}
\end{aligned}
\qquad (10.9)
$$

and the pressure p

$$
p = \frac{\partial F}{\partial n} n - F \qquad (10.10)
$$

appears as the Legendre transform of F and is expressed in terms of the chemical potential $\mu = \frac{\partial F}{\partial n}$.

Furthermore, by interpreting ζ as the specific Planck potential per molecule, one recovers for the entropy per molecule σ the expression given by statistical mechanics (8.6). This model is thus, in the non-dissipative limit, in agreement with the predictions of statistical mechanics. Introducing the 1-form

$$
H_\mu = hU_\mu \ , \text{ with } h = \frac{p+\rho}{\eta} \text{ [25]} \qquad (10.11)
$$

and its exterior derivative

$$
\Omega_{\lambda\mu} = \partial_\lambda H_\mu - \partial_\mu H_\lambda \qquad (10.12)
$$

we can replace the equations of motion (10.4) and (8.4) by

$$
\hat{\partial}_\mu N^\mu = 0 \ , \quad \Omega_{\mu\nu}\theta^\mu + \partial_\nu s = 0.
$$

This last equation shows that

$$
\delta S = 0 \text{ and } \delta_L\Omega = 0 \quad \text{for} \quad \delta X = \theta. \qquad (10.14)
$$

It follows that s is constant along each line of current and that Ω is an integral invariant of the field $X \leftrightarrow \theta$; its rank (4,2 or 0) is thus constant along each line of current, as well as the pseudo-scalar

$$
\pi = \mathrm{pf}(\Omega)\,\frac{\beta}{n} \qquad (10.15)
$$

where $\mathrm{pf}(\Omega)$ is the pfaffian of the form [26]. In general $\pi \neq 0$, Ω is of rank 4 and the sign of π defines an orientation of space. There exist also important classes of motions for which $\pi = 0$.

- The isentropic motions (those in which s keeps the same value along all lines of current); then (10.13) shows that $\theta \in \ker(\Omega)$; in general the rank of Ω is 2; the kernel of Ω defines a foliation whose leaves of dimension 2, can be interpreted as vortex lines carried away by the fluid. These notions are barotopic: there exists an equation of state, indexed by the value of s, obtained by eliminating β and n between p and ρ : the particular enthalpy h (10.11) coincides with the index defined by Lichnerowicz (15).

- Non isentropic motions in which rank (Ω) = 2 (it is sufficient that this be true at some arbitrary time); they make up the relativistic equivalent of the *oligotropic* motions of Casal (1); the "leaves" of Ω are described on the hypersurfaces $S = C^{st}$.

- The motions where $\Omega = 0$ (here again it is sufficient to verify this at some arbitrary time); (10.13) shows that they are isentropic; they constitute the *irrotational* notions, in the sense of Lichnerowi .z (15).

There exist solutions with discontinuities on a hypersurface Σ' of V_4 (*shock waves*); the conditions obtained by writing equations (10.4) in the sense of distributions read (27)

$$\begin{aligned} &N'^{\lambda} - N^{\lambda} \quad \text{is tangent to } \Sigma \\ &H'_{\lambda} - H_{\lambda} \quad \text{is normal to } \Sigma \\ &h'^2 - h^2 = [u'h'+uh][p'-p] \ ; \end{aligned} \tag{10.16}$$

if one adds that the discontinuity S'-S is positive, one obtains the shock equations of Rankine-Hugoniot in their relativistic form.

XI. WEAKLY DISSIPATIVE MOTIONS

The generating function ϕ can be put in general, taking into account condition (8.8), in the form

$$\begin{aligned} &\phi = \overset{\circ}{T}{}^{\mu\nu}\gamma_{\mu\nu} + \phi \\ &\overset{\circ}{T}{}^{\mu\nu} = \frac{\partial F}{\partial g_{\mu\nu}} \text{ and } \left.\frac{\partial \phi}{\partial \gamma_{\mu\nu}}\right|_{\gamma = 0} = 0 \end{aligned} \tag{11.1}$$

ϕ is called the dissipation function as one can show that

$$\hat{\text{div}}\ S = \frac{\partial\phi}{\partial\gamma_{\mu\nu}}\gamma_{\mu\nu}. \tag{11.2}$$

The quadratic approximation of ϕ in terms of γ is what is called the weakly dissipative approximation of the model

$$\phi = \frac{1}{2}[\lambda a^2+E_{ij}\chi^i\chi^j+F_{ij,lm}k^{ij}k^{lm}+2aB_i\chi^i+2aL_{ij}k^{ij}+2R_{i,jL}\chi^i k^{ji}] \tag{11.3}$$

(λ,E,F,B,L,R) are the dissipation coefficients; they are functions of (q,β,h) and total 55.

We recover the thermal conduction tensor E and viscosity tensor F; their components satisfy of course the Onsager symmetry relations :

$$\begin{aligned} E_{ij} &= E_{ji} \\ F_{ij,hl} &= F_{ji,hl} = F_{ij,lh} = F_{lh,ij}\ . \end{aligned} \tag{11.4}$$

To these coefficients, the model adds :

- λ, which we shall call the thermal susceptibility;
- the tensors B_i, L_{ij}, $R_{i,jk}$ which couple the effects of conduction and susceptibility, viscosity and susceptibility, conduction and viscosity , respectively.

Their components satisfy the symmetry relations

$$\begin{aligned} L_{ij} &= L_{ji} \\ R_{i,jk} &= R_{i,kj}\ . \end{aligned} \tag{11.5}$$

The reader is referred to (7) for the expression of the momentum-energy tensor and the equations of notion.

Furthermore, the limit $\lambda = 0$, $L = 0$, $R = 0$ and $B = 0$ lead in the Newtonian approximation to the Fourier heat equations and the Navier viscosity equations. Note that the conservation equations $\partial_\mu T^{\mu\nu} = 0$ lead, taking into account the convexity relations:

$$\begin{aligned} &\lambda > 0 \\ &\lambda E - B \otimes B > 0 \end{aligned} \tag{11.6}$$

a system of partial differential equations of the elliptic

kind. The elliptic character of these equations seems inevitable in as much as we have set ourselves at a macroscopic level. Taking into account derivatives of higher order of the kinematic variables (e.g. capillarity) would change of course the nature of the system of equations; one should not therefore give any fundamental interpretation to the fact that the system be elliptic or hyperbolic.

NOTES

(1) Unless mentioned explicitly, all functions considered in this paper will be taken to be C^{∞} ; in particular, $(\vec{r},t) \to \vec{F}$.

(2) i.e. $\sigma_V(dy)(\delta y) = 0, \forall\, \delta y$: this is the generalization of d'Alembert principle.

(3) A *density on a manifold* is a function f defined on the frames R and satisfying $f(RM) = f(R)|\det(M)|$ for any matrix M; on a symplectic manifold, there exists a density f_o, the *Liouville density* such that $f_o(R) = 1$ for any canonical frame. One can define the integral of a density with compact support on a manifold independently of any coordinate system; this allows one to identify each field of densities with a measure.

(4) By construction, ρ is a function defined on U; it is thus lifted on V through a first integral of the equations of motion; if one chooses an identification of the various phase spaces, this implies that ρ is a solution of Liouville's equation.

(5) A vector field F defined on a Hausdorff manifold U can be associated to the differential equation $\frac{dx}{ds} = F(x)$; the solution of this equation, which equals x_o for $S = 0$, is written as $\exp(sF)(x_o)$; if it exists $\forall\, x_o \in U$ and $\forall\, S \in \mathbb{R}$, F is said to be complete, then $S \mapsto \exp(sF)$ is a morphism of the group $(\mathbb{R},+)$ in the group of diffeomorphisms of U. If U is symplectic, and if $X \mapsto U$ is C^{∞} on U, then the *symplectic gradient* of the dynamical variable U is the vector field F defined by $\sigma(\delta x)(F(x)) = \delta U,\ \forall\, \delta$; the associated equation is the Hamilton equation; $\exp(sF)$ preserves the symplectic form σ and is therefore called a symplectomorphism.

(6) With the usual sign conventions, E must be replaced by -E.

(7) More precisely, these transformations (4.1) define a subgroup of the galilean group which is not an invariant subgroup.

(8) This is the Lie group, of dimension 10, generated by the isometries of $\mathbb{R}^3$: time-translations and galilean transformations $\vec{r} \to \vec{r} + \vec{a}t$, $\vec{v} \to \vec{v} + \vec{a}$.

(9) The group of isometries in Minkowski space, also of dimension 10.

(10) Q is an element of $\mathcal{G}^\star$, which generalizes the usual concept of "heat".

(11) In non-quantum statistical mechanics, the *spectrum* of a dynamical variable u in a statistical state μ is the *image* by $x \mapsto u$ of the probability law μ; it is a probability law of $\mathbb{R}$ (or of $\mathcal{G}^\star$ in the case of the moment).

(12) Another method which avoids certain topological difficulties, appeals to the *prequantization* algorithm (see (23)).

(13) The trial variable $x \mapsto \delta g$ being a covariant tensor field, T_x is called a distribution-tensor (**contravariant**).

(14) Z is said to be a Killing vector if exp(sZ) is an isometry $\forall$ s.

(15) The easiest way is to suppose that the support of T is *compact in space*, i.e. its intersection with any time slice $t_o \leqslant t \leqslant t_1$, being the line in an arbitrary Lorentz frame, is compact. This condition is satisfied by the T_x we have considered for a particle (provided it is not a tachyon !).

(16) The connected component of the neutral element of a Lie group is an invariant subgroup; the quotient by this subgroup is the *component group*. In the case of the Poincaré group, the component group is the Klein group (4 elements) which is abelian. Elements which "respect the orientation of time" make up the union of two components; they form thus an invariant subgroup called the orthochrone group.

(17) Usual entropy is the product by k of the quantity used here. Temperature units can always be chosen so as to make k=1.

(18) One will note that this argument relies implicitly on an approximation; on the one hand, one uses special relativity to construct the Poincaré moments; on the other hand one considers the $g_{\nu\rho}$ as variables as they give by differentiation the $T^{\nu\rho}$ through the Einstein equations. This approximation amounts to taking G to be small and to neglecting the gravitational self-interaction of the system; this is customary in thermodynamics.

(19) This current is defined by a 3-form on the reference manifold and lifts to space-time through a vector S by $vol_4(S)$, where vol_4 is the riemannian volume.

(20) $F = - Z/\beta$; Z is the specific Planck potential ; $\theta^\mu = \beta U^\mu$ $U^\mu U_\mu = 1$.

(21) ϕ must be understood as a function with density value.

(22) covariant, in the sense of densities, is the exact term.

(23) ϕ and F are still to be taken in the sense of densities.

(24) Definition which generalizes beyond equilibrium formula (8.1) established for Gibbs states.

(25) Because of the relativistic equivalence between specific mass and specific energy (c=1), h can be interpreted as the *enthalpy per particle*.

(26) This pfaffian is defined by $\frac{1}{2}\,\Omega \wedge \Omega = pf(\Omega)\,vol$, where vol represents the riemannian volume form defined through an orientation of V_4. π is the relativistic equivalent of the vortex potential of Ertel (5,6).

(27) Dasked variables are taken after the shock.

REFERENCES

(1) P. Casal : Principes variationnels en fluide compressible et en magnétodynamique des fluides. Journ. de Mécanique, 5, 2 (1966), pp. 149.

(2) B. Carter : Perfect fluid and magnetic field conservation laws in the theory of black holes accretion rings. DAF. Observatoire de Paris, Meudon (France).

(3) B. Carter : Regular and anomalous heat conduction. Journ. relativistes 1976, Université Libre de Bruxelles.

(4) A. Einstein : The meaning of relativity. Methuen, London (1922).

(5) H. Ertel : Meteorol. Ztschr., 59 (1942), pp. 277-385.

(6) H.L. Gavriline, MM. Zaslavskii : Dauk. Ak. Nauk SSSR, 192 1 (1970), pp. 48.

(7) P. Iglesias : Essai de thermodynamique rationnelle des milieux continus, Ann. de l'Institut H. Poincaré, vol. XXXIV. A paraître.

(8) W. Israel : Non-stationary irreversible thermodynamics : a causal relativistic theory, OAP. 444, Caltech (1976).

(9) W. Israel : Extract from "General relativity", papers in honour of J.L. Synge, Clarendon Press, Oxford (1972).

(10) W. Israel : The dynamics of polarisation, General relativity and gravitation, vol. 9, 5, (1978).

(11) W. Israel, J.M. Stewart : Thermodynamics of non-stationary and Transient effects in relativistic gas. Phys. Lett. 58 4, (1976).

(12) W. Israel, J.M. Stewart : Transient relativistic thermodynamics and kinetic theory. Ann. of Phys., 118, 2(1979).

(13) A.I. Khinchin : Mathematical foundations of statistical mechanics. Dover, New York (1949).

(14) L. Landau, E. Lifchitz : Statistical physics. Addison-Wesley (1969). Fluid Mechanics. Addison-Wesley (1958).

(15) A. Lichnerowicz : Théories relativistes de la gravitation et de l'électromagnétisme. Masson, Paris (1955).

(16) J.J. Moreau : Loi de l'élasticité en grande déformation. Séminaire d'Analyse convexe. Montpellier (1979).

(17) Noll : The foundation of mechanics and thermodynamics. Springer (1979).

(18) Sedov : Mécanique des milieux continus. Editions de Moscou.

(19) J.M. Souriau : Définition covariante des équilibres thermodynamiques. Supp. Nuovo Cimento, 1, I.4. (1966), pp. 203-216.

(20) J.M. Souriau : Structure des systèmes dynamiques. Dunod. Paris (1969).

(21) J.M. Souriau : Modèle de particule à spin dans le champ électromagnétique et gravitationnel. Ann. Inst. H. Poincaré. XX. 4 (1974), p. 315-364.

(22) J.M. Souriau : Thermodynamique relativiste des fluides. Rend. Sem. Mat., Univers. Politechn. Torino, 35 (1976-1977), p. 21-34.

(23) J.M. Souriau : Structure of dynamical systems. North-Holland. Amsterdam (à paraître).

(24) J.M. Souriau : Géométrie et relativité. Hermann (1964).

(25) J.L. Synge : The relativistic gas. Amsterdam (1957).

(26) C. Vallée : Thermodynamique relativiste des milieux continus. Institut de Mécanique des fluides. Poitiers. A paraître.

(27) G.N. Watson : A treatise of the theory of Bessel functions (1944).

DEFORMATIONS OF ALGEBRAS ASSOCIATED WITH A SYMPLECTIC MANIFOLD

A. Lichnerowicz

Collège de France

It is possible to give a complete description of Classical Mechanics in terms of symplectic geometry and Poisson brackets. It is the essential of the hamiltonian formalism. In a common program with Flato, D. Sternheimer and J. Vey and other scientists (Fronsdal, Arnal, M. Cahen, S. Gutt, M. de Wilde) we have studied properties and applications of the deformations of the trivial associative algebra and of the Poisson Lie algebra associated with a symplectic manifold. Such deformations give a new approach for Quantum Mechanics; this approach has been developped in other papers ((1),(2)). In this lecture, I will give recent results concerning the existence and the equivalence of associative deformations (or $\star_\nu$-products).

I. THE ALGEBRAS

a) Let (W,F) be a smooth connected symplectic manifold of dimension 2n and fundamental 2-form F. We denote by $b_k(W)$ the $k^{\underline{th}}$ Betti number of W for the cohomology with unrestricted supports. All the elements are supposed to be C^∞. We put for simplicity $N = C^\infty(W;\mathbb{R})$. Let $\mu : TW \to T^\star W$ be the isomorphism of vector bundles defined by $\mu(X) = - i(X)F$ (where i(.) is the inner product);

M. Cahen et al. (eds.), Differential Geometry and Mathematical Physics, 69–83.

this isomorphism can be extended to tensors in a natural way. We denote by Λ the anti-symmetric contravariant 2-tensor $\mu^{-1}(F)$.

A *symplectic vector field* is a vector field X such that $\mathcal{L}(X)\, F = 0$ (where $\mathcal{L}$ is the Lie derivative); it is an infinitesimal automorphism of the structure; X is symplectic iff the 1-form $\mu(X)$ is closed. We denote by L the (infinite dimensional) Lie algebra of the symplectic vector fields. If $X, Y \in L$, we have

(1-1) $\quad \mu([X,Y]) = d\, i(\Lambda)(\mu(X) \wedge \mu(Y))$

Let $L^\star$ be the subspace of L defined by the converse images of the exact 1-forms ($X_u = \mu^{-1}(du) : u \in N$). An element of $L^\star$ is a *hamiltonian* vector field. Consider the commutator ideal [L,L] of L; it is well-known (Arnold, myself) that $[L,L] = L^\star$ and $\dim L/L^\star = b_1(W)$. We are led to introduce the Poisson bracket :

(1-2) $\quad \{u,v\} = i(\Lambda)(du \wedge dv) = \mathcal{L}(X_u)v = P(u,v) \quad (u,v \in N)$

where the Poisson operator P is a bidifferential operator of order 1 for each argument, null on the constant; P defines on N a structure of Lie algebra and (N,P) is *the Poisson Lie algebra* of the manifold. Since $X_{\{u,v\}} = [X_u,X_v]$ we have a homomorphism of (N,P) onto $L^\star$.

b) The space N admits the following two algebraic structures:

1) a structure of *associative algebra* given by the usual product of functions which is here commutative)

2) a structure of *Lie algebra* given by the Poisson bracket. The Poisson bracket defines derivations of the product. It is natural to study if it is possible *to deform* these two algebraic laws in a suitable and consistent way, so that we obtain a model isomorphic to the conventional Quantum Mechanics. The answer is positive.

II. HOCHSCHILD COHOMOLOGY AND CHEVALLEY COHOMOLOGY

a) Derivations and deformations of an *associative* algebra arise from a same cohomology, the so-called *Hochschild cohomology*. Let W be an arbitrary differentiable manifold and $(N = C^\infty(W;\mathbb{R}),.)$ the associative algebra defined by the product of functions.

A p-cochain C ($p \geq 1$) of (N,.) is a p-linear map of N^p into N. The Hochschild coboundary of the p-cochain C is the (p+1)-cochain $\tilde{\partial}C$ defined by :

$$\tilde{\partial}C(u_o,..,u_p) = u_o C(u_1,..,u_p) - C(u_o u_1, u_2,..,u_p) + C(u_o, u_1 u_2,..u_p) \tag{2-1}$$
$$+(-1)^p C(u_o,u_1,..,u_{p-1}u_p) + (-1)^{p+1} C(u_o,..,u_{p-1})u_p \ .$$

We have $\tilde{\partial}^2 = 0$. A 1-cocycle of (N,.) is a derivation of the algebra and so is defined by a vector field. A p-cochain C is said to be d-*differential* ($d \geq 0$) if it is defined by a multidifferential operator of maximum order d in each argument. If T is an endomorphism (1-cochain) of N which is (d+1)-differential, $\tilde{\partial}T$ is d-differential. Conversely

<u>Proposition</u>. *If* T *is an endomorphism of* N *such that* $C = \tilde{\partial}T$ *is* d-*differential* ($d \geq 0$), T *is* (d+1)-*differential itself*.

If $\tilde{\partial}T$ is null on the constants, it is the same for T. Let $\tilde{H}^p(N;N)$ be the pth Hochschild cohomology space for the differential Hochschild cohomology; J. Vey (3) has proved by means of results of Gelfand the following

<u>Theorem (Vey)</u>. $\tilde{H}^p(N;N)$ *is isomorphic to the space of the antisymmetric contravariant* p-*tensors of* W.

Such a tensor defines an alternate multidifferential operator of order 1 and can be identified with this operator. In dimensions 2 and 3, I have given an elementary proof of the Vey theorem and M. Cahen, S. Gutt and M. de Wilde have extended this result to the *local* cohomology. In general, we consider here cochains which are *null on the constants* for which the Hochschild cohomology is not changed.

b) In a symmetric way, derivations and deformations of a Lie algebra arise from a same cohomology, the so-called *Chevalley cohomology* of the Lie algebra corresponding to the adjoint representation. Let (W,F) be a symplectic manifold and (N,P) the corresponding Lie algebra. A Chevalley p-cochain C ($p \geq 0$) is here an *alternate* p-linear map of N^p into N, the 0-cochains being identified with the elements of N. The Chevalley coboundary ∂ is classicaly defined by the formula :

$$\partial C(u_o,..u_p) = \varepsilon^{\lambda_o..\lambda_p}_{o..p}\left(\frac{1}{p!}\{u_{\lambda_o}, C(u_{\lambda_1},..,u_{\lambda_p})\}\right.$$
$$\left. - \frac{1}{2(p-1)!} C(\{u_{\lambda_o}, u_{\lambda_1}\}, u_{\lambda_2},..,u_{\lambda_p})\right)$$

where $u_\lambda \in N$ and where ε is the Kronecker skewsymmetrization indicator. A 1-cocycle of (N,P) is a derivation of the Lie algebra, an exact 1-cocycle being an inner derivation. For the d-differential character of a cochain C, we have definitions similar to the definitions concerning the Hochschild cohomology; but we suppose here $d \geq 1$: if C is d-differential, ∂C is also d-differential. Conversely I have proved :

Proposition. *If* C *is an exact* d-*differential* ($d \geq 1$) *Chevalley two-cocycle of* (N,P), *there is a differential operator of order* d *such that* $C = \partial T$.

Avez and myself (4) have determined all the derivations of (N,P) without a priori differentiability assumption. In particular the derivations which are null on the constants are given by $\mathcal{L}(X)u$, where X is a symplectic vector field. We denote by $H^p(N;N)$ the pth Chevalley cohomology space for the differential Chevalley cohomology *with cochains which are null on the constants.*

III. FORMAL DEFORMATIONS

I will recall now and extend the main elements of the theory of Gerstenhaber (5) concerning the deformations of the algebraic structures, in particular the associative algebras.

a) Let $E(N;\nu)$ be the space of the formal functions of $\nu \in \mathbb{C}$ with coefficients in N; ν is said to be the deformation parameter. Consider a bilinear map $N \times N \to E(N;\nu)$ which gives the formal series:

(3-1) $$u \star_\nu v = \sum_{r=0}^{\infty} \nu^r C_r(u,v) = uv + \sum_{r=1}^{\infty} \nu^r C_r(u,v)$$

where the C_r are differential 2-cochains of (N,.). These cochains can be extended to $E(N;\nu)$ in a natural way. We say that (3-1) defines *a formal deformation* of (N,.) if the associativity identity holds formally. If such is the case, (3-1) defines on $E(N;\nu)$ a structure of *formal associative algebra*. If the map (3-1) is arbitrary, we have for $u,v,w \in N$:

$$(u \star_\nu v) \star_\nu w - u \star_\nu (v \star_\nu w) = \sum_{t=1}^{\infty} \nu^t \tilde{D}_t(u, v, w) \tag{3-2}$$

where $\tilde{D}_t$ is the 3-cochain :

$$\tilde{D}_t(u,v,w) = \sum_{r+s=t} (C_r(C_s(u,v),w) - C_r(u,C_s(v,w))) \quad (r,s \geqslant 0)$$

We are led to put :

$$\tilde{E}_t(u,v,w) = \sum_{r+s=t} (C_r(C_s(u,v),w) - C_r(u,C_s(v,w))) \quad (r,s \geqslant 1)$$

and we have the identity :

$$\tilde{D}_t \equiv \tilde{E}_t - \tilde{\partial} C_t$$

If (3-1) is limited at the order q, we have a deformation of order q if the associativity identity is satisfied up to the order (q+1). If such is the case, $\tilde{E}_{q+1}$ is automatically a 3-cocycle of (N,.). We can find a 2-cochain C_{q+1} satisfying $\tilde{D}_{q+1} = \tilde{E}_{q+1} - \tilde{C}_{q+1}$ iff $\tilde{E}_{q+1}$ is exact; $\tilde{E}_{q+1}$ defines a cohomology class, element of $\tilde{H}^3(N;N)$, which is *the obstruction at the order* (q+1) *to the construction of a deformation*. A deformation of order 1 is called infinitesimal. We have $\tilde{E}_1 = 0$ and so only $\tilde{\partial} C_1 = 0$, that is C_1 is a 2-cocycle of (N,.).

b) Consider a formal series in ν:

$$T_\nu = \sum_{s=0}^{\infty} \nu^s T_s = Id_N + \sum_{s=1}^{\infty} \nu^s T_s \tag{3-3}$$

where the T_s $(s \geqslant 1)$ are endomorphisms of N; T_ν acts naturally on $E(N;\nu)$. Consider also another bilinear map $N \times N \to E(N;\nu)$ corresponding to the formal series

$$u \star'_\nu v = uv + \sum_{r=1}^{\infty} \nu^r C'_r(u,v) \tag{3-4}$$

where the C'_r are differential 2-cochains again. Suppose that (3-3) , (3-4) are such that we have formally the identity :

(3-5) $T_\nu (u \star'_\nu v) = T_\nu u \star_\nu T_\nu v$

We can prove by means of universal formulas :

Proposition. *The deformation* (3-1) *of* (N,.) *being given, each formal series* (3-3) *, where the* T_s *are necessarily differential operators, generates a unique bilinear map* (3-4) *satisfying* (3-5); *this map is a new deformation which is said to be equivalent to* (3-1). *In particular a deformation is called trivial if it is equivalent to identity deformation* ($C_r = 0$, $r \geq 1$).

According to (3-5), T is said to transform $\star'_\nu$ in $\star_\nu$. If two deformations are equivalent at the order q, there appears a 2-cocycle such that its cohomology class, element of $\tilde{H}^2(N;N)$, is *the obstruction to the equivalence for order* (q+1).

In particular two infinitesimal deformations defined by the 2-cocycles C_1 and C'_1 are equivalent iff $(C'_1 - C_1)$ is exact.

c) Let $E(N;\lambda)$ be the space of the formal functions of $\lambda \in \mathbb{C}$ with coefficients in N. A deformation of the Poisson Lie algebra (N,P) is defined by an *alternate* bilinear map $N \times N \to E(N;\lambda)$ given by

$$(3\text{-}6) \qquad [u,v]_\lambda = P(u,v) + \sum_{r=1}^{\infty} \lambda^r C_{2r+1}(u,v)$$

where the C_{2r+1} are differential 2-cochains of (N,P) such that the Jacobi identity holds formally. The Chevalley cohomology plays exactly the same role for the deformations of Lie algebras as the Hochschild cohomology for the deformations of associative algebras.

IV. THE STAR-PRODUCTS

a) A Hochschild 2-cochain C(u,v) is said to be *even* if it is symmetric in u,v, odd if it is antisymmetric. Similar definition for a Hochschild 3-cochain E(u,v,w) with respect to the symmetry relatively to u,w. If C is a 2-cochain, $\tilde{\partial} C$ is odd for an even C, even for an odd C. It follows from the Vey theorem that if C is an *even* (resp. *odd*) Hochschild 2-cocycle, it is *exact* (resp. *non exact* and 1-*differential*). Let E be a Hochschild 3-cocycle; if E is even it is exact and $E = \tilde{\partial} C^{(o)}$, if E is odd, $E = T + \tilde{\partial} C^{(e)}$, where $C^{(o)}$ (resp. $C^{(e)}$) is odd (resp. even) and where T is given by a skewsymmetric 3-tensor.

b) On the symplectic manifold (W,Λ) P corresponding to the structure 2-tensor Λ is *a non exact Hochschild 2-cocycle.* We consider only in the following the associative deformations of the form :

$$(4\text{-}1) \qquad u \star_\nu v = uv + \nu P(u,v) + \sum_{r=2}^{\infty} \nu^r C_r(u,v)$$

where the C_r satisfy the following assumption :

1) *the* C_r *are null on the constants*

2) C_r *is even if* r *is even, odd if* r *is odd.*

If such is the case, we say that (4-1) defines a $\star_\nu$-product (or *star-product*) on (W,Λ). We have :

$$u \star_\nu 1 = 1 \star_\nu u = u \qquad u \star_{-\nu} v = v \star_\nu u$$

A star-product generates by skewsymmetrization a formal Lie algebra (3-6) with $\lambda = \nu^2$.

$$(4\text{-}2) \qquad [u,v]_\lambda = (2\nu)^{-1}(u \star_\nu v - v \star_\nu u)$$

I have proved the following :

Uniqueness theorem (6). *If a formal Lie algebra* (3-6) *is generated by a star-product, this star-product is unique.*

c) Let $\star_\nu$, $\star'_\nu$ be two equivalent star-product, T_ν transforming $\star'_\nu$ in $\star_\nu$; $A_\nu = T_{-\nu}(T_\nu)^{-1}$ defines an automorphism of $\star_\nu$ satisfying $A_{-\nu} = (A_\nu)^{-1}$. I have proved by induction that there is a unique automorphism B_ν of $\star_\nu$ (with a trivial main part) such that $A_\nu = B_\nu^2$ and satisfying $B_{-\nu} = (B_\nu)^{-1}$ changing B_ν in $B_{-\nu}$, we see that T_ν admits a unique decomposition

$$(4\text{-}3) \qquad T_\nu = B_\nu T_\nu^{(e)}$$

where $T_\nu^{(e)}$ is even in ν and transforms $\star'_\nu$ in $\star_\nu$ and B_ν is an automorphism of $\star_\nu$ satisfying $B_{-\nu} = (B_\nu)^{-1}$. Therefore two star products which are equivalent are even equivalent with respect to ν . It follows that two formal Lie algebras generated by two equivalent star-products are equivalent with respect to $\lambda = \nu^2$.

d) It is easy to see that, for a star-product, $\tilde{E}_t$ is even if t is odd, odd if t is even. Considering together the two cohomologies and using parity considerations, Neroslavsky and Vlassov have proved an important existence theorem (7) :

Existence theorem (N. V.). *Each symplectic manifold* (W,F) *such that* $b_3(W) = 0$ *admits star-products.*

V. THE VEY STAR-PRODUCTS

a) A *symplectic connection* Γ is a linear connection without torsion such that $\nabla F = 0$, where ∇ is the operator of covariant differentiation defined by Γ. It is easy to see that a symplectic manifold admits infinitely many symplectic connections; the difference between two symplectic connections is deduced from an arbitrary symmetric covariant 3-tensor.

Let Γ be a symplectic connection on (W,Λ). We set $P^o(u,v)=u.v$, $P^1 = P$ and introduce the bidifferential operators of maximum order r in each argument, defined for each domain U of an arbitrary chart (x^i) $(i,j,\ldots = 1,\ldots,2n)$ by the following expression

$$(5\text{-}1)\quad P^r(u,v)|_U = P^r_\Gamma(u,v)|_U = \Lambda^{i_1 j_1}\ldots\Lambda^{i_r j_r}\,\nabla_{i_1\ldots i_r}u\nabla_{j_1\ldots j_r}v \qquad (u,v \in N)$$

If Γ *is flat,* $\exp(\nu P)$ defines a star-product (the so-called *Moyal product*).

b) This situation can be generalized in the following way. Let Γ be an arbitrary symplectic connection; P and P^2_Γ define always a $\star_\nu$-product at order 2. If $u,v \in N$, denote by $\mathcal{L}(X_u)$ the symmetric covariant 3-tensor deduced from the Lie derivative of Γ by the hamiltonian vector field X_u. The 2-cochain S^3_Γ given by :

$$(5\text{-}2)\quad S^3_\Gamma(u,v)|_U = \Lambda^{i_1 j_1}\Lambda^{i_2 j_2}\Lambda^{i_3 j_3}(\mathcal{L}(X_u)\Gamma)_{i_1 i_2 i_3}(\mathcal{L}(X_v)\Gamma)_{j_1 j_2 j_3}$$

is *a non-exact Chevalley 2-cocycle* $(\partial S^3_\Gamma = 0)$ which admits the same principal symbol as P^3

$$uv + \nu P(u,v) + (\nu^2/2!)\,P^2(u,v) + (\nu^3/3!)\,S^3_\Gamma(u,v)$$

defines a star-product at order 3. The Chevalley cohomology 2-

class β of S^3_Γ is independent of Γ and is *an invariant of the symplectic structure*. We can prove

Proposition. *The second space* $H^2(N;N)$ *of Chevalley cohomology admits as generators the class* β *and the classes defined by the images by* μ^{-1} *of the closed 2-form of* W.

We see that $H^2(N;N)$ depends upon β and the de Rham cohomology of W in dimension 2. We have a similar result for $H^3(N;N)$ (8). In particular each 1-differential Chevalley cocycle is the image by μ^{-1} of a closed form; it is exact if the form is exact.

c) Introduce now the following notation : we denote by Q^r a bidifferential operator of maximum order r in each argument, null on the constants, satisfying the parity assumption and such that its principal symbol coincides with the principal symbol of P^r_Γ . We take in particular $Q^o(u,v) = uv$, $Q^1 = P$. We introduce the following :

Definition. A *Vey star-product* is a star-product of the form :

$$(5\text{-}3) \qquad u \star_\nu v = \sum_{r=0}^{\infty} (\nu^r/r!)\, Q^r(u,v)$$

A *Vey Lie algebra* is a formal Lie algebra given by a bracket of the form :

$$(5\text{-}4) \qquad [u,v]_\lambda = \sum_{r=0}^{\infty} (\lambda^r/(2r+1)!)\, Q^{2r+1}(u,v)$$

My viewpoint differs from the view-point of Vey who considers essentially the Lie algebras. Consider a Vey $\star_\nu$-product at the order 2; we can show that there is a unique symplectic connection Γ such that

$$(5\text{-}5) \qquad Q^2 = P^2_\Gamma + \tilde{\partial} H$$

where H is a differential *operator of maximum order* 2. Suppose that Q^3 is a Chevalley 2-cocycle ($\partial Q^3 = 0$). I have proved similarly that there is a unique connection Γ such that :

$$(5\text{-}6) \qquad \partial Q^3 = S^3_\Gamma + T + 3 \quad \partial H$$

where H is a differential operator of maximum order 2 and T a 2-tensor image by μ^{-1} of a closed 2-form; P, $Q^2/2$, $Q^3/6$ give a $\star_\nu$-product at order 3 iff *the symplectic connections and the operators*

H *of* (5-5), (5-6) *coincide.*

d) I have proved by means of a long study of the bidifferential types the following :

Theorem. *Each star-product of* (W,F) *is equivalent to a Vey star-product.*

We can deduce from this theorem and from the general existence theorem that each symplectic manifold (W,F) such that $b_3(W) = 0$ admits a Vey star-product. Moreover, using the recursion process of Neroslavsky-Vlassov, we can show that we can choose $Q^3 \in \beta$. We see by skewsymmetrization that each symplectic manifold (W,F) such that $b_3(W) = 0$ admits Vey Lie algebra such that $Q^3 \in \beta$. This last result has been obtained directly by Jacques Vey by a completely different and difficult method.

VI. THE CASE WHERE $b_2(W) = 0$

a) For a symplectic manifold (W,F), such that $b_2(W) = 0$, the 2-form F is exact and $H^2(N,N)$ admits only β as generator. We will show :

Theorem. *For a symplectic manifold* (W,F) *such that* $b_2(W) = 0$, *all the star-products are equivalent.*

Consider on (W,F) two star-products :

$$(6\text{-}1) \qquad u *_\nu v = uv + \nu P(u,v) + \sum_{r=2}^{\infty} \nu^r C_r(u,v)$$

and

$$(6\text{-}2) \qquad u *'_\nu v = uv + \nu P(u,v) + \sum_{r=2}^{\infty} \nu^r C'_r(u,v)$$

We proceed by induction and suppose that, by transformation of (6-1) we have obtained $C'_r = C_r$ for $r \leqslant 2q-1$ (with $q \geqslant 1$); the even Hochschild 2-cocycle $C'_{2q} - C_{2q}$ is exact so that :

$$C'_{2q} = C_{2q} + \tilde{\partial}A$$

where A is a differential operator. Transforming (6-1) by means of $T_\nu = Id + \nu^{2q}A$, we obtain $C_r = C'_r$ for $r = 1,.., 2q$. If such is the case, we have $C'_{2q+1} - C_{2q+1} = T$, where T is an antisymmetric 2-tensor image by μ^{-1} of a closed 2-form, that is exact; T is also an exact Chevalley 2-cocycle and there is a vector Z such that

$T = \partial\mathcal{L}(Z)$. The equivalence given by $T_\nu = Id. + \nu^{2q}\,\mathcal{L}(Z)$ preserves the C_r for $r \leqslant 2q$ and transforms C_{2q+1} in $C_{2q+1} + T = C'_{2q+1}$. We see that all the star-products are equivalent.

b) Up to now, we have considered only equivalence operators T_ν which are defined by differential operators which are null on the constants. For the study of the formal Lie algebras, we are led to introduce equivalence operators which do not satisfy this condition, but for which the differential operators are constant on the constants. We say that we have *a weak equivalence*. Similarly, we consider associative deformations satisfying the parity assumption and for which $C_1 = P$ and *the cochains of odd rank* are null on the constants, *the cochains of even rank* being non necessarily null on the constants. Such deformation is called *a weak star-product*. It is possible to show that *a weak star-product has the following form :*

(6-3) $\quad u \star_\nu v = k_{\nu^2}(u \,\hat{\star}_\nu v) \qquad k_{\nu^2} = 1 + \sum_{r=1}^{\infty} \nu^{2r}\, k_{2r} \quad (k_{2r} \in \mathbb{R})$

where $\hat{\star}_\nu$ *is a star-product.*

Weak star-product generate by skewsymmetrization formal Lie algebras *of the form* :

(6-4) $\quad [u,v]_\lambda = P(u,v) + \sum_{r=1}^{\infty} \lambda^r\, C_{2r+1}\,(u,v)$

where the C_{2r+1} *are null on the constants.* For such weak star-products, the uniqueness theorem is valid again. S. Gutt has proved :

<u>Proposition</u>. *If* $b_2(W) = 0$, *all the formal Lie algebras of the form* (6-4) *and such that* $C_3 = Q^3/3!$ *are weakly equivalent.*

VII. FORMAL LIE ALGEBRAS GENERATED BY A WEAK STAR-PRODUCT.

a) Consider, on an arbitrary symplectic manifold (W,F), a formal Lie algebra (6-4) *such that* $C_3 = Q^3/3!$ (with $\lambda = \nu^2$). Let U be a contractible domain of W ($b_2(U) = 0$) and introduce the restriction of the considered Lie algebra to U. This Lie algebra on $(U,F|_U)$ is weakly equivalent to the Moyal Lie algebra on U generated by a Moyal product defined by means of a Darboux chart of F of domain U. Our Lie algebra $\{C_{2r+1}|_U\}$ is thus generated by a weak star-product $\{C_{2r+1}|_U\}$ on U. We set :

$$C_{2r(U)}(u,v) = \bar{C}_{2r(U)}(u,v) + k_{2r(U)}uv \quad (k_{2r(U)} \in \mathbb{R})$$

where $\bar{C}_{2r(U)}$ is null on the constants. Consider two contractible domains U, V of W, with $U \cap V \neq \emptyset$. We deduce from the uniqueness theorem that on $U \cap V$:

$$k_{2r(U)} = k_{2r(V)} = k_{2r} \qquad \bar{C}_{2r(U)} = \bar{C}_{2r(V)}$$

Therefore there exist bidifferential $\bar{C}_{2r}$, null on the constants, such that $\bar{C}_{2r(U)} = \bar{C}_{2r}|_U$. The $\bar{C}_{2r}$, k_{2r}, C_{2r+1} define on W a weak star-product which generates the given Lie algebra.

It is possible to prove that each weak star-product is equivalent to a weak Vey star-product. It follows that the formal Lie algebra (6-4) with $C_3 = Q^3/3!$ is equivalent to a Vey Lie algebra. It is the same if $C^3_3 = Q^3/3! + A$, where A is a differential operator null on the constants. We deduce from this argument :

Theorem. *A formal Lie algebra* (6-4) *is equivalent to a Vey Lie algebra if and only if the corresponding Chevalley* 2*-cocycle* C_3 *is cohomologous to a* 2*-cocycle of the form* $Q^3/3!$

b) We have seen that such a Lie algebra is generated by a weak star-product. Conversely consider a formal Lie algebra (6-4) which is generated by a weak star-product. Each weak star-product being equivalent to a weak star-product, our Lie algebra is equivalent to a Vey Lie algebra. We have :

Theorem. *A formal Lie algebra* (6-4) *is generated by a weak star-product if and only if it is equivalent to a Vey Lie algebra.*

It is easy to see that, a Vey Lie algebra being given, there is a unique Vey Lie algebra, deduced from this by product by a constant k_{ν^2}, which is generated by a Vey star-product.

VIII AUTOMORPHISMS OF A STAR-PRODUCT.

a) Let (W,F) be a symplectic manifold admitting a $\star_\nu$-product. Consider an automorphism of the space $E(N;\nu)$:

$$A_\nu = A_o + \sum_{s=1}^{\infty} \nu^s A_s$$

where A_o is an automorphism of the space N and the A_s ($s \geqslant 1$) are endomorphisms of N; A_ν *is an automorphism of the* $\star_\nu$*-product if*

(8-1) $$A_\nu(u \star_\nu v) = A_\nu u \star_\nu A_\nu v \qquad (u,v \in N)$$

If A_o = Id., we say that A has *a trivial main part*. Let $\hat{E}(N;\nu)$ be the subset of $E(N;\nu)$ defined by the elements $a_\nu = \Sigma \nu^s a_s$ such that $a_o > 0$ on W. Each element a_ν of $\hat{E}(N;\nu)$ admits an inverse element (denoted by $a_\nu^{(\star)-1}$) in the sense of the $\star_\nu$-product. An *inner automorphism* of $\star_\nu$ is given by $u \to a_\nu \star_\nu u \star_\nu a^{(\star)-1}$ and has a trivial main part.

b) Let *Symp*(W,F) be the group of all the symplectomorphisms of (W,F) and $Symp_c$(W,F) its differentiably arcwise connected component of the identity. We denote by $\mathcal{D}$ the space of the differential operators which are null on the constants. Suppose that (W,F) admits a formal Lie algebra (6-3). I have proved (6) that, for for each $\sigma \in Symp_c$ (W,F, there is an automorphism of this formal Lie algebra which has the form:

(8-2) $$A_\nu = (Id + \sum_{s=1}^{\infty} \nu^{2s} B_{2s})\sigma^\star \qquad (B_{2s} \in \mathcal{D}; \lambda = \nu^2)$$

c) It is possible to show that each automorphism of a star-product has necessarily the form

(8-3) $$A_\nu = (Id + \sum_{s=1}^{\infty} \nu^s B_s)\sigma^\star \qquad (B_s \in \mathcal{D})$$

where σ is a symplectomorphism. Conversely, we can prove by means of the uniqueness theorem:

<u>Proposition</u>. *The group* $Aut_t(\star_\nu)$ *of the automorphisms of the* $\star_\nu$*-product admitting a trivial main part coincides with the group of the automorphisms of the corresponding Lie algebra which have the form*

$$A_\nu = Id + \sum_{s=1}^{\infty} \nu^s A_s \qquad (A_s \in \mathcal{D})$$

If $b_1(W) = 0$,, $Aut_t(\star_\nu) = Aut_i(\star_\nu)$, *where* $Aut_i(\star_\nu)$ *is the group of inner automorphisms.*

We can deduce from this proposition by means of a study of the automorphisms which are even in ν:

Theorem. *The group* $Aut(\star_\nu)$ *of the automorphisms of the* $\star_\nu$*-product coincides with the group of the automorphisms of the corresponding Lie algebra which have the form* (8-3).

The group $Aut(\star_\nu)/Aut_t(\star_\nu)$ is thus isomorphic to a subgroup K of $Symp(W,F)$; it follows from b that $Symp_c(W,F) \subset K$. We can think that $K = Symp(W,F)$

d) Consider *a Vey* $\star_\nu$*-product.* We have seen that there is a unique symplectic connection Γ such that $Q^2 = P^2_\Gamma + \tilde{\partial} H$ (H differential operator of order ≤ 2). The study of the automorphisms independent of ν of the Vey $\star_\nu$-product shows the following:

Proposition. *The group of the automorphisms independent of* ν *of a Vey* $\star_\nu$*-product is a closed subgroup of the group of the affine symplectomorphisms for the corresponding connection* Γ. *It follows that this group is a finite dimensional Lie group.*

REFERENCES

(1) M. Flato, A. Lichnerowicz, D. Sternheimer. C.R. Acad. Sci. Paris, 283 A, (1976), pp. 19-24.

(2) F. Bayen, M. Flato, C. Fronsdal, A. Lichnerowicz, D. Sternheimer. Lett. in Math. Phys. 1, (1977), pp. 521-530; Ann.of Physics 111, (1978), PP. 61-132.

(3) J. Vey. Comm. Math. Helv. 50, (1975), pp. 421-454.

(4) A. Avez and A. Lichnerowicz. C.R. Acad. Sci. Paris, 275 A, (1972) pp. 11-24.

(5) M. Gerstenhaber. Ann. of Math. 79, (1964), pp. 59-103.

(6) A. Lichnerowicz. C.R. Ac. Sci. Paris, 286 A, (1978), pp. 49-53; Ann. di Matematica t 123, (1980), pp. 287-330.

(7) M. Neroslavsky and A. Vlasov. C.R. Acad. Sci. Paris I, t 292, (1981), pp. 71-75.

(8) S. Gutt. Lett. in Math. Phys. 3(1979), pp. 297-310; Ann. Inst. H. Poincaré 27, (1980).

(9) A. Lichnerowicz. C.R. Acad. Sci. Paris 287 A, (1978) pp. 1029-1033; Lett. in Math. Phys. 3, (1979), pp. 495-502.

(10) M. Cahen, S. Gutt and M. De Wilde. Local cohomology of the algebra of C^{∞}-functions on a connected manifold; Letters in Math. Phys. 4(1980), pp. 157-167.

DIFFERENTIAL DEFORMATIONS WITH CONSTANT COEFFICIENTS

G. Patissier

University of Lyon I

I. INTRODUCTION

Let W be $\mathbb{R}^q$ and $E(W,\lambda)$ be the space of formal series with coefficients in $C^\infty(W)$. A formal bi-differential operator P_λ is a bilinear map $P_\lambda : C^\infty(W)^2 \to E(W,\lambda)$ with $P_\lambda = \sum_{s=0}^{\infty} \lambda^s P_s$ where $P_s : C^\infty(W)^2 \to C^\infty(W)$ is a usual bi-differential operator.

Let $\mathcal{S}(W)$ be the space of rapidly decreasing functions, if $u \in \mathcal{S}(W)$ we denote by $\hat{u}$ the Fourier transformation of u.P_λ can be formally written

$$P_\lambda(u,v) = \frac{1}{(2\pi)^{2q}} \int e^{i<x,\eta+\xi>} a(\lambda,x,\eta,\xi)\, \hat{u}(\eta)\, \hat{v}(\xi)\, d\eta\, d\xi \tag{1.0}$$

with

$$a(\lambda,x,\eta,\xi) = \sum_{s=0}^{\infty} \lambda^s a_s(x,\eta,\xi),$$ where $a_s(x,\eta,\xi)$ is a polynomial in the (η,ξ) variables with coefficients in $C^\infty(W)$; $a(\lambda,x,\eta,\xi)$ is called the symbol of P_λ .

1.1. Examples. (i) If $P_o(u,v) = u.v$ (or equivalently $a_o(x,\eta,\xi)=1$) we obtain a formal deformation of the usual associative algebra defined on $C^\infty(W)$ by $(u,v) \to u.v$. In such a situation we let $u \star v = P_\lambda(u,v)$.

(ii) We denote by $\{u,v\}$ the Poisson bracket defined by

M. Cahen et al. (eds.), Differential Geometry and Mathematical Physics, 85–94.

$$\{u,v\} = \sum_{i=1}^{n} (\partial_{n+i}u\ \partial_i v - \partial_i u\ \partial_{n+i} v)$$

where $u,v \in C^\infty(\mathbb{R}^{2n})$. For $(\eta_1,\ldots,\eta_{2n}) \in \mathbb{R}^{2n}$ let $F(\xi,\eta) = \sum_{i=1}^{n} (\xi^{n+i}\eta^i - \xi^i\ \eta^{n+i})$ be the canonical symplectic 2-form over $\mathbb{R}^{2n}$, $-F(\xi,\eta)$ is the symbol of Poisson bracket.

In order to emphasize the polynomial aspects of deformation theory we have avoided the use of cohomology. We substitute cohomological conditions, used in the general Poisson manifolds, ((1), (3),(7),(9)), for algebraic conditions over symbols

$$\sum_{s=0}^{\infty} \lambda^s a_s(x,\eta,\xi).$$

Generally the coefficients a_s will be assumed to be functions of the variables (ξ,η) only (that is we have formal operators with constant coefficients). With this hypothesis, we obtain simple conditions over symbols. Nevertheless, there are interesting examples in this most restrictive situation. Notice there is no hypothesis neither on symmetry or antisymmetry of operators P_λ neither on order.

II. DEFORMATIONS OF ASSOCIATIVE ALGEBRA $C^\infty(W)$.

Let $\star$ be a formal bi-differential operator with constant coefficients and symbol $a(\lambda,\eta,\xi)$. Compute the symbols of formal tri-differential operators $(u,v,w) \to u \star (v \star w)$ and $(u,v,w) \to (u \star v) \star w$, we obtain

2.1. <u>Proposition</u>. $a(\lambda,\eta,\xi)$ *defines an associative deformation of the usual associative algebra if and only if there exists a sequence of polynomials* $b_s(\eta,\xi)$, $s \geqslant 1$, *such that*

(i) $a(\lambda,\eta,\xi) = \exp\left[\sum_{s=1}^{\infty} \lambda^s b_s(\eta,\xi)\right]$ (2.1.0)

(ii) *for every integer* s

$$b_s(\eta,\xi) + b_s(\eta+\xi,\zeta) = b_s(\eta,\xi+\zeta) + b_s(\xi,\zeta)$$

2.2. <u>Examples</u>. (i) Let $\phi_k : W^2 \to \mathbb{C}$, $k \geqslant 1$, be a sequence of $\mathbb{R}$-bilinear maps and $C \in \mathbb{C}[[\lambda]]$ a formal series such that $C(0) = 0$.

The symbol

$$a(\lambda,\eta,\xi) = \exp\,[c(\lambda) + \sum_{k=1}^{\infty} \lambda^k \phi_k(\eta,\xi)]$$

satisfies 2.0

(ii) Let $R_s(\eta)$, $\eta \in W$, be a sequence of polynomials, we set

$$b_s(\eta,\xi) = R_s(\eta+\xi) - R_s(\eta) - R_s(\xi).$$

For $d^\circ R_s > 2$, b_s is not bilinear. Then for $q = 1$, we obtain all the polynomials satisfying (2.0) (ii).

2.3. Applications. We set $W = \mathbb{R}^{2n}$, let $a(x,\xi) \in \mathcal{S}'(W)$ be a tempered distribution. For i = 1, 2, 3 we define the linear operators

$$Op_i : \mathcal{S}'(W) \to \mathcal{L}(\mathcal{S}(\mathbb{R}^n), \mathcal{S}'(\mathbb{R}^n))$$

by

$$Op_i(a).u(x) = \frac{1}{(2\pi)^n} \int e^{i<x,y>} a[\theta_1(x,y),\xi]\, u(y)dy\, d\xi .$$

where

$\theta_1(x,y) = x$ (classical theory of pseudo-differential operators (4)).

or $\theta_2(x,y) = \frac{x+y}{2}$, (Weyl transform (5), (10)),

or $\theta_3(x,y) = y$ (we obtain the pseudo-differential operators studied by Kohn-Nirenberg (6).

Let a be a symbol of P.D.O. theory (see (4), (5), (6), (10), for example) we can prove that

$$O_{p_i}(a) \circ Op_i(b) = O_{p_i}(a \star b)$$

$$a \star b = a \star_\lambda b/_{\lambda=1} \qquad (2.3.0)$$

where $\star_\lambda$ is a differential deformation with constant coefficients.

For i=1 the symbol of $\star_\lambda$ is

$$\exp\,[-i\lambda \sum_{j=1}^{n} \eta^{n+j} \xi^j]$$

(in this case, the $\star$-product is denoted by #).

For i=2, the symbol of $\star_\lambda$ is

$$\exp\,[-i\lambda \sum_{j=1}^{n} \eta^j \xi^{n+j}] .$$

(where the relation (2.3.0) is written in the sense of the asymptotic calculus of pseudo differential operators, cf. (4),(5)).

2.4. Let

$$f(Z) = 1 + \sum_{k=1}^{\infty} \alpha_k Z^k$$

be a formal series with complex coefficients. The following generalizes a proposition proved for Moyal product in (1).

Proposition. *Let* $\exp b(\lambda,\eta,\xi)$ *be a symbol of an associative deformation of usual associative algebra* $C^{\infty}(W)$. *The* $\star$*-product defined by* $f(b(\lambda,\eta,\xi)$ *is associative if and only if* $f(Z) = \exp(KZ)$, $K \in C$.

2.5. Remarks. (i) Let $\phi : W \times W \to \mathbb{R}$ be an antisymmetric bilinear map. We denote by $\star_\lambda$ the $\star$-product defined by the symbol $a(\lambda,\eta,\xi) = \exp [\lambda i \ \phi \ (\eta,\xi)]$. For $\lambda=1$, the relation (1.0) defines a continous bilinear map $\star : S(W) \times S(W) \to S(W)$, and as Moyal product we prove that

$$\int u\star v(x)dx = \int u(x)v(x)dx.$$

(ii) We consider the equation $u_\lambda \star_\lambda x = \omega_\lambda$ for the formal series $x \in E(W,\lambda)$, where u_λ, $\omega_\lambda \in E(W,\lambda)$.

if $u_o(x) \neq 0$, $\forall y \in W$, this equation can be formally solved. We have

$$x = \sum_{s=o}^{\infty} \lambda^s x_s$$ with inductive relations like the relations obtained, for $\#$ product, in the calculus of the parametrix of an elliptic pseudo-differential operator (4).

III. EQUIVALENCE OF FORMAL BI-DIFFERENTIAL OPERATORS.

Let P_λ and Q_λ be two formal bi-differential operators with constant coefficients and respective symbols $p(\lambda,\eta,\xi)$ and $q(\lambda,\eta,\xi)$. Remind that (1)

3.1. Definition. P_λ *and* Q_λ *are said to be equivalent if there exists a formal differential operator* $T_\lambda = Id + \sum_{s=1}^{\infty} \lambda^s T_s$ (with any coefficients) such that

$$\forall u, v \in C^{\infty}(W), \ T_\lambda \ [P_\lambda \ (u,v)] = Q_\lambda[T_\lambda \ (u), \ T_\lambda(v)] \ .$$

3.2. Proposition. *An equivalence* T_λ *is a formal differential operator with constant coefficients if and only if its symbol* $a(\lambda,\eta,\xi)$ *satisfies*

$$p(\lambda,\eta,\xi)\ a(\lambda,\eta+\xi) = q(\lambda,\eta,\xi)\ a(\lambda,\eta)\ a(\lambda,\xi).$$

3.3. Examples. The three deformations defined by operators Op_i $(i=1,2,3)$ in section 2.3 are equivalent. For example let $\star$ be the Moyal product, we have

$$T_{2,\lambda}\ (a \# b) = T_{2,\lambda}(a)\star\ T_{2,\lambda}(b)$$

where $T_{2,\lambda}$ is the formal operator of symbol $\exp(-\frac{\lambda i}{2} < x,\xi>)$. This is related to the fact that the three algebras of pseudo-differential operators defined by operators Op_i are the same provided that we limit operators Op_i to a good subalgebra of $C^\infty(W)$ (5).

3.4. Corollary. *For* $P_\lambda \neq 0$ *the automorphisms with constant coefficients of* P_λ . *(for* $P_\lambda = Q_\lambda$*) are defined by symbols*

$$a(\lambda,\eta) = \exp\ [\ \sum_{k=1}^{\infty}\ \lambda^k\ b_k(\eta)]$$

where $b_k : W \to \mathbb{C}$ *are* $\mathbb{R}$*-linear maps.*

IV. TRIVIAL STAR-PRODUCTS WITH CONSTANT COEFFICIENTS.

Let $\star$ be a (associative or no) deformation of usual product of $C^\infty(W)$ with symbol $\exp\ [a(\lambda,\eta,\xi)]$ where

$$a(\lambda,\eta,\xi) = \sum_{s=1}^{\infty}\ \lambda^s\ a_s(\eta,\xi).$$

4.1. Definitions (1) . (i) $\star$ *is a trivial product if it is equivalent to usual product of* $C^\infty(W)$, *that is there exists a formal differential operator* T_λ *(with any coefficients but with* $T_o = Id$ *such that*

$$T_\lambda\ (u\star v) = T_\lambda(u)T_\lambda(v). \qquad (4.1.1)$$

(ii) $\star$ *is a* k*-trivial product if the identity* 4.1.1 *is limited to the order* k, *that is there exists* k *differential operators such that the identity* 4.1.1 *is satisfied up to the order* k+1 *when we set*

$$T_\lambda = Id + \sum_{s=1}^{k}\ \lambda^s\ T_s$$

4.2. Proposition. $\star$ *is a* k- *trivial product if and only if*

$$a_h(\eta,\xi) = c_h + \theta_h(\eta,\xi) \qquad 1 \leq h \leq k,$$

where c_h *is a constant map and* $\theta_h : W^2 \to C$ *a* $\mathbb{R}$ *bilinear map.*

Then for $\phi_h(\eta) = c_h - \frac{1}{2}\theta_h(\eta,\eta)$ *the symbol of operator* T_λ *can be chosen as*

$$\exp \sum_{h=1}^{k} \lambda^h \phi_h(\eta) \quad .$$

4.3. Examples. The Moyal product and the # product (cf. 2.3) are not trivial.

V. INVARIANCE OF STAR-PRODUCTS.

Let W be $\mathbb{R}^{2n}$, suppose that deformation is not necessarily with constant coefficients, denote by $a(\lambda,\eta,\xi) = 1 + \sum_{s=1}^{\infty} \lambda^s a_s(x,\eta,\xi)$ its symbol.

5.1. Definition (1). *Let* ϕ *be a function,* $\star$ *is said to be* ϕ*-invariant if*

$$\{\phi,u\star v\} = \{\phi,u\} \star v + u\star\{\phi,v\} \ , \quad \forall\ u\ ,\ v \in C^\infty(W) \ . \tag{5.1.1}$$

Denote by INV($\star$) the set of such ϕ, INV($\star$) is a Lie algebra of the Poisson algebra $C^\infty(W)$.

Example. For Moyal product, INV($\star$) is the Lie algebra of polynomials of degree 2 (1).

5.2. Proposition. *We set*

$$p(\lambda,x,\eta,\zeta,\xi) = a(\lambda,x,\eta,\zeta)F(\xi,\eta+\zeta) - a(\lambda,x,\xi+\eta,\zeta)F(\xi,\eta) - a(\lambda,x,\eta+\xi,\zeta)F(\xi,\zeta).$$

Denote by $P_\lambda(\eta,\zeta,\frac{\partial}{\partial x})$ the formal differential operator defined by the symbol $p(\lambda,x,\zeta,\xi)$ (with respect to (λ,x,ξ)). This operator is $\star$-invariant if and only if

$$P_\lambda(\eta,\zeta,\frac{\partial}{\partial x})\phi = \{\phi(.),a(\lambda,.,\eta,\zeta)\} \ , \ \forall \eta \ , \ \zeta \in W \ .$$

5.3. Corollary. $\phi \in$ INV(#) *if and only if* $\phi(p,q) = \chi(p,q) + \theta(p,q)$ *for a polynomial* χ *of degree less than* 1 *and a bilinear map* $\theta : \mathbb{R}^n \times \mathbb{R}^n \to \mathbb{C}$.

5.4. Remark. Denote by H_n the Lie algebra of polynomials of degree less than 1 (Heisenberg algebra). We have $H_n \subset$ INV($\star$) if and only if $\star$ is a deformation with constant coefficients.

VI. DEFORMATIONS OF THE POISSON ALGEBRA $C^{\infty}(W)$.

Let $[\,,]_{\lambda}$ be a formal bi-differential operator with constant coefficients and symbol $a(\lambda,\eta,\xi)$ such that $a(0,\eta,\xi) = -F(\eta,\xi)$.

6.1. Proposition. $(E(W,\lambda)\ ,\ [\,,]_{\lambda})$ *is a Lie algebra if and only if*

$$\begin{aligned} &\text{(i)} \quad a(\lambda,\eta,\xi) = -a(\lambda,\xi,\eta) \\ &\text{(ii)} \quad \oint a(\lambda,\eta,\xi)a(\lambda,\eta+\xi,\zeta) = 0. \end{aligned} \tag{6.1.1}$$

where $\oint$ *is the summation over cyclic permutations of* (η,ξ,ζ).

6.2. Examples (i) Let $\star$ be the Moyal product, we set

$$[u,v]_{\lambda} = \frac{1}{2\lambda i}(u\star v - v\star u)$$

$[u,v]_{\lambda}$ is the Moyal-Vey bracket, this is a deformation of Poisson bracket defined by the symbol $-\frac{1}{\lambda}\sin\lambda F(\eta,\xi)$ which satisfies the identities (6.1.1).

(ii) We set $[u,v]_{\lambda} = \frac{-1}{\lambda i}(u \# v - v \# u)$; we obtain a deformation of Poisson bracket which furnishes the symbol of commutator of two pseudo-differential operators.

(iii) Let $G_1 : W \times W \to \mathbb{C}$, $i = 1,2\ldots$ a sequence of antisymmetric $\mathbb{R}$-bilinear maps.

We set

$$a_2(\lambda,\eta,\xi) = -F(\eta,\xi) + \sum_{k=1}^{\infty} \lambda^k G_k(\eta,\xi).$$

Thus, we obtain 1-differentiable deformations with constant coefficients (3).

VII. TRIVIALITY OF DEFORMATIONS OF POISSON ALGEBRA $C^{\infty}(W)$.

Let $[u,v]_{\lambda}$ be a deformation of Poisson bracket with constant coefficients and symbol

$$a(\lambda,\eta,\xi) = -F(\eta,\xi) + \sum_{k=1}^{\infty} \lambda^k a_k(\eta,\xi).$$

7.1. Definitions (1). (i) $[u,v]_{\lambda}$ *is a trivial deformation if it is equivalent to Poisson bracket, that is there exists a formal differential operator* T_{λ} *such that* $T_0 = \mathrm{Id}$ *and*

$$T_\lambda([u,v]_\lambda) = \{T_\lambda u, T_\lambda v\} \qquad (7.1.1)$$

(ii) $[u,v]$ *is a* k*-trivial deformation if the identity* (7.1.1) *is satisfied up to the order* k+1 , *that is there exists* k *differential operators* $T_1,..,T_k$ *such that the identity* (7.1.1) *is verified up to the order* k+1 *with*

$$T_\lambda = Id + \sum_{h=1}^{k} \lambda^h T_h .$$

7.2. Proposition. *Let* T_λ *be a formal differential operator with symbol* $b(\lambda,x,\eta)$, $[u,v]_\lambda$ *is a* k*-trivial deformation by interposition of* T_λ *is and only if we have the following identity*

$$a(\lambda,\eta,\xi)\, b(\lambda,x,\eta+\xi) = -F(\eta,\xi)\, b(\lambda,x,\eta)\, b(\lambda,x,\xi) + \{b(\lambda,x,\eta),b(\lambda,x,\xi)\}_x$$
$$+ ib(\lambda,x,\eta)F[\eta,\mathrm{grad}_x b(\lambda,x,\xi)] - ib(\lambda,x,\xi)F[\xi,\mathrm{grad}_x b(\lambda,x,\eta)] . \qquad (7.2.0)$$

7.3. Corollary. $[u,v]_\lambda = \frac{-1}{i\lambda}(u \# v - v \# u)$ *is a 1-trivial deformation but it is not 2-trivial.*

In the study of triviality of deformation of associative algebra we can limit ourselves to consider operator T_λ with constant coefficient only (cf 4).
But we go to prove in the following, that is not the same for deformations of Poisson bracket.

7.4. Definition. *The deformation* $[u,v]_\lambda$ *will said to be* C*-trivial up to the order* k+1 *if* $[u,v]_\lambda$ *is trivial and if the operators* $T_1,\dots T_k$ *in the definition* (7.1) *can be chosen with constant coefficients.*

7.5. Examples. Let $G_h : W \times W \to \mathbb{C}$ be a sequence or antisymmetric bilinear maps. We set

$$a(\lambda,\eta,\xi) = -F(\eta,\xi) + \sum_{h=1}^{\infty} \lambda^h G_h(\eta,\xi)$$

the deformation defined by $a(\lambda,\eta,\xi)$ is C-trivial up to the order k+1 if and only if there exists k complex numbers : $K_1,\dots,K_k$ such that $G_h = K_h F$ for $1 \leqslant h \leqslant k$.

7.6. Corollary. C*-triviality is not involved by triviality.*

Indeed, we set $a(\lambda,\eta,\xi) = -F(\eta,\xi) + \lambda G_1(\eta,\xi)$ where

$G_1(\eta,\xi) = \eta^1 \xi^{n+1} - \eta^{n+1} \xi^1$ and dim $W = 2n > 2$. From after the previous example, $a(\lambda,\eta,\xi)$ defines a Poisson deformation which is not C-trivial up to the order 2, but this is a 1-trivial deformation for identity (7.2.0) is satisfied with

$$b_1(x,\eta) = ix_1\eta^1 \quad \text{where } x = (x_1,\ldots,x_{2n}) \in W.$$

REFERENCES.

(1) F. Bayen, M. Flato, C. Fronsdal, A. Lichnerowicz, D. Sternheimer : Deformation theory and quantification. I. Deformation of symplectic structures. Ann. of Phys., t. III (1978), pp. 61-110.

(2) C. Chevalley, S. Eilenberg : Cohomology theory of Lie groups and Lie algebras. Trans. Amer. Soc., t. 63 (1948), pp.85-124.

(3) M. Flato, A. Lichnerowicz, D. Sternheimer: Déformations 1-différentiables des algèbres de Lie attachées à une variété symplectique ou de contact. Compositio Mathematica, Vol. 31, Fasc. 1(1975), pp. 47-82.

(4) L. Hormander : Pseudo-differential operators and hypoelliptic equations. Amer. Math. Soc., Symp. in pure Math., t. 10(1966).

(5) L. Hormander : The Weyl calculus of pseudo-differential operators. Comm. on pure and appl. Math., Vol. XXXII (1979), p. 359-443.

(6) J.J. Kohn, L. Nirenberg : An algebra of pseudo-differential operators. Comm. on pure and appl. Math., Vol. XVIII (1965), p. 269-305.

(7) A. Lichnerowicz : Dérivations et automorphismes pour des algèbres formelles associées par déformation à une variété symplectique. C.R. Acad. Sci. Paris 286 (1978).

(8) G. Patissier : Déformations différentiables à coefficients constants et produits de Moyal généralisés. Ann. Inst. Henri Poincaré, Section A, Physique Théorique, Vol. XXX, n°4 (1979), p. 275-293.

(9) J. Vey : Déformation du crochet de Poisson sur une variété

symplectique. Comm. Math. Helvetia 50 (1975), p. 421-454.

(10) A. Voros : An algebra of pseudo-differential operators and the asymptotics of quantum mechanics. Journal of Functional Analysis, Vol. 29, n°1 (1978),pp. 104-132.

Lectures given at the
Meeting of the Belgian Contact Group
on Differential Geometry
held at Leuven, February 6-8, 1981

SOME ASPECTS OF THEORETICAL PHYSICS RELATING TO HARMONIC MAPS

P. Baird

I.H.E.S., Bures-sur-Yvette, France

I. INTRODUCTION

Let (M,g) be a compact Riemannian manifold without boundary, and consider an action which, for simplicity, we suppose has the form

$$I(\phi) = \int_M L(j^1(\phi))\, dx \ , \qquad (1.1)$$

where ϕ is a section of a Riemannian fiber bundle $\pi : E \to M$, and $L : J^1(E) \to \mathbb{R}$ possibly depends on the metric of E.

Example 1.2. Let $\phi : (M,g) \to (N,h)$ be a smooth map, $E = M \times N$. Then ϕ can be regarded as a section of E, and $L(j^1(\phi)) = e_\phi = \frac{1}{2} \| d\phi \|^2$ is the Hilbert-Schmidt norm of $d\phi$.

In general one looks for extremals of (1.1) with respect to variations of the section ϕ, and one finds that ϕ is an extremal if and only if ϕ satisfies the Euler-Lagrange equations

$$\varepsilon L(\phi) = 0 \quad . \qquad (1.3)$$

In example 1.2, the extremals are the harmonic maps and the Euler-Lagrange equation for ϕ is

$$\tau_\phi = \operatorname{div}(d\phi) = 0 \quad . \qquad (1.4)$$

Suppose now we vary the metric g. If g(t) is a smooth 1-parameter family of metrics, $g(0) = g$, then

M. Cahen et al. (eds.), Differential Geometry and Mathematical Physics, 97–104.

$\delta g = \partial g/\partial t\big|_{t=0} \in \Gamma(\odot^2 T^{*}M)$, and we find

$$\frac{dI(\phi)}{dt}\bigg|_{t=0} = \int_M < S_\phi, \delta_g > dx \quad , \tag{1.5}$$

for a fixed section ϕ, where $S_\phi \in \Gamma(\odot^2 T^{*}M)$ and $< \, , >$ is the metric induced on $\odot^2 T^{*}M$ from g. In example (1.2) we find

$$S_\phi = e_\phi g - \phi^{*}h \quad . \tag{1.6}$$

Suppose X is a smooth vector field on M, and let L_X denote Lie derivation with respect to X, then

$$0 = \int_M L_X(L\ dx)$$

$$= \int_M < \varepsilon L(\phi), d\phi(X)\ dx + \int_M \operatorname{div} S_\phi(X)\ dx \quad .$$

Since X is arbitrary, we conclude that

$$< \varepsilon L(\phi), d\phi > + \operatorname{div} S_\phi = 0 \quad . \tag{1.7}$$

In particular, if ϕ is an extremal of (1.1) so that (1.3) is satisfied, then $\operatorname{div} S_\phi = 0$. We note that in the case of example (1.2) we have

$$< \tau_\phi, d_\phi > + \operatorname{div} S_\phi = 0 \quad . \tag{1.8}$$

Such divergence free symmetric tensors are well-known and important in relativity theory, where they in some sense model the matter distribution in a space-time model. For if $(S_{ij})_{1\leq i,j\leq 3}$ are the space components of a symmetric 2-tensor, then S_{ij} is the i-component of a force associated to the j-vector (the "stress" acting on a unit area orthogonal to j). Since force is a time rate of change of momentum, this represents the rate of flow of the i-component of momentum through a unit area orthogonal to j. We introduce the time components of S, to obtain the 4-tensor $(S_{ij})_{0\leq i,j\leq 3}$, where now S_{io} represents energy flow, and S_{oo} energy density. That div S = 0 and S be symmetric, corresponds to conservation laws. For if X is a Killing vector field, then

$$\operatorname{div}(S(X)) = 0 \quad ,$$

which corresponds to conservation of momentum if X is spacelike, and energy if X is timelike.

The fundamental equations of A. Einstein (2) are

$$\frac{1}{2}R^M g - \mathrm{Ricci}^M = S \quad , \tag{1.9}$$

relating the curvature and the matter of space. Shortly after Einstein gave these equations, Hilbert gave the above method of deriving stress-energy tensors from a variational principle (4). Following a suggestion of Taub (1963), we use the stress-energy tensor in the form (1.6) to study harmonic maps (1).

II. APPLICATIONS

1. (a) Suppose $\phi: (M,g) \to (N,h)$ is conformal with $\phi^{*}h = \rho g$. Then (1.6) gives

$$S_\phi = \frac{1}{2}\,\rho(m-2)g \quad .$$

Thus if $m = 2$, $S_\phi = 0$, and if $m > 2$ and ϕ is harmonic, (1.8) implies ρ is constant. Conversely, if $S_\phi = 0$, then $0 = \mathrm{trace}\, S_\phi = \frac{1}{2}(m-2)\, e_\phi$, which implies $m = 2$ and $\phi^{*}h = e_\phi g$, i.e. ϕ is conformal. So we have

$$(S_\phi \equiv 0) \text{ if and only if } (\dim M = 2 \text{ and } \phi \text{ is conformal}) \tag{2.1}$$

2. Let $\phi : (M,g) \to (N,h)$ be a map of a Riemann surface to a Riemannian manifold. Choose isothermal coordinates $(z,\bar{z})$ on M so that $g : \sigma^2\, dz\, d\bar{z}$. Then the complexified tangent space $T^{\mathbb{C}}M = \mathbb{C}\,\{\frac{\partial}{\partial z}, \frac{\partial}{\partial \bar{z}}\}$ decomposes as

$$T^{\mathbb{C}}M = T'M + T''M \quad ,$$

where $T'M = \mathbb{C}\,\{\frac{\partial}{\partial z}\}$, $T''M = \mathbb{C}\,\{\frac{\partial}{\partial \bar{z}}\}$. We can associate the dual spaces $T^{*}{}'M = \mathbb{C}\,\{dz\}$, $T^{*}{}''M = \mathbb{C}\,\{d\bar{z}\}$, and S_ϕ can be decomposed into types

$$S_\phi = S_\phi^{(2,0)} + S_\phi^{(1,1)} + S_\phi^{(0,2)} \quad , \tag{2.3}$$

and we find that $S_\phi^{(1,1)} = 0$ and

$$S^{(2,0)} = \overline{S^{(0,2)}} = h^{\mathbb{C}}(d\phi(\frac{\partial}{\partial z}), d\phi(\frac{\partial}{\partial z}))dz^2 \quad ,$$

where $h^{\mathbb{C}}$ is the symmetric bilinear extension of h to $T^{\mathbb{C}}N$. Thus

$$S_\phi = S_\phi^{(2,0)} + S_\phi^{(0,2)} \quad . \tag{2.4}$$

Furthermore $\mathrm{div}\, S_\phi = 0$ if and only if $S_\phi^{(2,0)}$ is a holomorphic quadratic differential on M. In particular, if M is homeomorphic to S^2, such differentials don't exist, and $S_\phi = 0$, which by (2.1) implies ϕ is conformal.

3. (a) Let $\phi: (M,g) \to (\mathbb{R}^N, \langle\ ,\ \rangle)$ be an isometric immersion of an m-dimensional Riemannian manifold into a Euclidean space. Then we can define the Gauss map

$$\gamma : (M,g) \to (G(m,\mathbb{R}^N),k)$$

of M into the Grassmannian of m-planes in $\mathbb{R}^N$. Then for $x \in M$

$$d\gamma_x : T_xM \to T_{\gamma(x)}G(m,\mathbb{R}^N)$$
$$\simeq \mathrm{Hom}(T_xM, N_xM) \quad ,$$

where N_xM is the normal space to M in $\mathbb{R}^N$ at x. Thus $d\gamma$ can be regarded as a section of $\odot^2 T^* M \otimes NM$, and in fact can be identified with the second fundamental form $\nabla d\phi$ of ϕ (6) .

Thus $\nabla d\gamma = \nabla(\nabla d\phi)$, and

$$\begin{aligned}\mathrm{Trace}\ \nabla d\gamma &= \mathrm{Trace}\ \nabla(\nabla d\phi)\\ &= \nabla(\mathrm{Trace}\ \nabla d\phi)\\ &= \nabla(m \times \text{mean curv. vector field}),\end{aligned}$$

(we can commute Trace and ∇ since we are mapping into $\mathbb{R}^N$). Thus M has constant mean curvature in $\mathbb{R}^N$ if and only if the Gauss map is harmonic. The same arguments apply if $\mathbb{R}^N$ is relaced by the standard sphere or hyperbolic space.

We calculate S_γ using a formula of Obata for $\gamma^* k$ (5), and obtain

$$S_\gamma = \frac{|\tau\phi|^2 - R^M + cm(m-1)}{2} . g - \beta_\phi . \tau_\phi + \mathrm{Ricci}^M - c(m-1)g \quad , \tag{2.5}$$

where β_ϕ is the second fundamental form of ϕ, and c is the constant curvature of the range manifold. In particular if ϕ is a minimal immersion, so $\tau_\phi = 0$, then

$$S_\gamma = \frac{-R^M}{2} g + \mathrm{Ricci}^M + \frac{cg}{2}(m-1)(m-2) \quad . \tag{2.6}$$

This has a striking resemblance to (1.9), and if γ is conformal; ϕ represents a solution of Einstein's vacuum equations.

Example . The Veroneze map $\phi : S^2 \to S^4$ defined by

$$\phi(x,y,z) = \left(xy, xz, yz, \frac{x^2-y^2}{2}, \frac{x^2+y^2-2y^2}{2\sqrt{3}}\right),$$

isometrically immerses S^2 into S^4 as $\mathbb{R}P^2$.

(b) Much of the above applies to the case when g has signature $(-1,1,\overbrace{\ldots\ldots}^{r},1)$, and $\mathbb{R}^N$ is equipped with an indefinite metric of signature (p,q) say. In particular γ is defined as a map into a suitable homogeneous space with a natural indefinite metric k:

$$\gamma : M \to \frac{O(P,q)}{O(1,r) \times O(p-1,q-r)} \quad ,$$

where $P \geqslant 1$ and $q \geqslant r$. Furthermore $d\gamma = \nabla\, d\phi$, and $\phi(M)$ has constant mean curvature implies γ is harmonic.

The stress-energy tensor of γ has the same form as (2.5)

$$S_\gamma = \frac{|\tau_\phi|^2 - R^M}{2}\, g - \beta_\phi.\tau_\phi + \text{Ricci}^M \tag{2.7}$$

In certain circumstances one can consider maps into spaces of non-zero curvature - this is best illustrated by an example.

Example. Compactified Minkowski space M can be realized as a 4-dimensional projective quadric Q in $\mathbb{R}\,P^5$, defined by the equation

$$-x_1^2 - x_2^2 + x_3^2 + x_4^2 + x_5^2 + x_6^2 = 0 \quad ,$$

in the homogeneous coordinates $[x_1,\ldots,x_6]$ for $\mathbb{R}P^5$. This quadric is minimal in $\mathbb{R}P^5$ with respect to the standard Riemannian metric on $\mathbb{R}P^5$. The Gauss map γ can be viewed as a map from $\mathbb{Q}$ to $G(5,\mathbb{R}^6)$, defined by sending each point x to the 5-plane determined by the tangent space to Q at x and the line in $\mathbb{R}^6$ defining x. Thus $\gamma : Q \to \mathbb{R}P^5$, and in fact

$$\gamma[x_1,\ldots,x_6] = [-x_1,-x_2,x_3,\ldots,x_6] \quad , \tag{2.8}$$

so γ is a harmonic and isometric immersion.

At each point x of Q, the intersection of the tangent space to Q at x with Q defines the light cone structure in TQ, and one can in a natural way replace the Riemannian metric with a Lorentzian metric with the same light cones, to obtain M. This Lorentz metric arizes by restricting the indefinite metric $ds^2 = -dx_1^2 - dx_2^2 + dx_3^2 + \ldots dx_6^2$ on $\mathbb{R}^6$ to Q. The new Gauss map still defined by (2.8), maps Q into

$\mathbb{R}P^5$ with a non-standard metric, and is still harmonic and isometric. Thus

$$S_\gamma = -\frac{R^M}{2} g + \text{Ricci}^M = 0 \quad ,$$

and M satisfies Einstein's vacuum equations.

4. Higher order Gauss maps

Let $\phi: (M,g) \to \mathbb{R}^N$ be an isometric immersion of a Riemannian manifold into standard Euclidean space, and let $\alpha : \mathbb{R} \to M$ be a smooth curve with $\alpha(0) = x$. The p'th osculating space to α at x is the span of

$$\left(\frac{D^j \alpha(0)}{ds^j}\right)_{1 \leqslant j \leqslant p} \quad ,$$

where $\frac{D}{ds}$ represents covariant derivation in $\mathbb{R}^N$ along α .
The p'th osculating space $T_x^{(p)}M$ to M at x is the span of all the p'th osculating spaces at x to all curves α in M through x .
Let $r_p(x)$ denote the dimension of $T_x^{(p)}M$ - this may vary with x.
Let $U \subset M$ be an open subset on which r_p is constant for all p.
The p'th Gauss map

$$\gamma^{(p)} : (U,g) \to (G(r_p, \mathbb{R}^N) k^p) \quad ,$$

is defined by $\gamma^{(p)}(x) = T_x^{(p)}M$ for all $x \in U$. Let $N_x^{(p)}M$ denote the orthogonal complement of $T_x^{(p)}M$ in $T_x^{(p+1)}M$. Then the (p+1)'th fundamental form is defined by $\beta^{(p+1)} = \pi^{(p)} (\nabla^p \, d\phi)$, where $\nabla^p(d\phi) \in \Gamma(\otimes^{p+1} T^* M \otimes \phi^{-1} \, T\mathbb{R}^N)$ and $\pi^{(p)}(x) : T_{\phi(x)}\mathbb{R}^N \to N_x^{(p)}M$ is projection. Then we have

Proposition 2.9. (3) *The derivative* $d\gamma^{(p)}$ *is equal to* $\beta^{(p+1)}$.

We note that in particular if $\gamma^{(p)}$ is homothetic then $\nabla d\gamma^{(p)} = d\gamma^{(p+1)}$.

Suppose M is a Riemann surface endowed with isothermal coordinates $z,\bar{z}$ such that $g = \rho^2 dz \, d\bar{z}$. The complex connection splits as $\nabla^{\mathbb{C}} = \nabla' + \nabla''$, and define ∂ and $\bar{\partial}$ by

$$\partial\phi = \frac{\partial\phi}{\partial z} = \phi_z$$

$$\bar{\partial}\phi = \frac{\partial\phi}{\partial\bar{z}} = \phi_{\bar{z}} \quad .$$

Then ∂ and $\bar{\partial}$ commute, whereas ∇' and ∇'' do not commute in general.

Given any harmonic map $\psi : (M,g) \to (N,h)$ between Riemannian manifolds, we consider the following formula for Stress-Energy tensors

$$\frac{1}{2} \text{Trace } \nabla^2 S_\psi = -S_{A(\psi)} + S_{B(\psi)} + S_{C(\psi)} \quad , \qquad (2.10)$$

where for a square matrix M,

$$S_M = \frac{1}{2}(\text{Trace } M)g - \odot M \quad ,$$

and in local coordinates

$$A(\psi)_{ij} = R^N_{\alpha\beta\gamma\delta} \psi^\beta_\ell \psi^\delta_n g^{\ell n} \psi^\alpha_i \psi^\gamma_j$$

$$B(\psi)_{ij} = R^{Mk}_{i} (\psi^* h)_{kj}$$

$$C(\psi)_{ij} = \psi^\alpha_{i,m} \psi^\beta_{j,n} g^{mn} h_{\alpha\beta} \quad ,$$

and for the map we write $S_\psi = S_{\psi^* h}$. Note that if the matrix M is proportional to the metric g, and dim M = 2, then $S_M = 0$. Thus if ϕ is as above

$$\frac{1}{2} \text{Trace } \nabla^2 S_\phi = -S_{A(\phi)} + S_{B(\phi)} + S_{C(\phi)} \quad .$$

Now $A(\phi)$ vanishes since $\mathbb{R}^N$ has zero curvature, S_ϕ vanishes since ϕ is an isometric immersion (from (2.1)), $S_{B(\phi)} = -\text{Ricci}^M + \frac{1}{2}R^M g = 0$ and $S_{C(\phi)} = S_\gamma$ since $\nabla d\phi = d\gamma$. Thus $S_\gamma = 0$ and by (2.1) γ is conformal.

<u>Proposition 2.11</u>. *If* $\gamma^{(p)}$ *is homothetic and harmonic, then* $S_{\gamma(p+1)} = 0$, *i.e.* $[\gamma^{(p+1)^*} k^{\mathbb{C}(p)}]^{(2,0)} = 0$, *where* $k^{\mathbb{C}(p)}$ *is the complex extension of the metric* $k^{(p)}$ *on* $G(r_p, \mathbb{R}^N)$.

<u>Proof</u> : Since $\gamma^{(p)}$ is homothetic $S_{C(\gamma^{(p)})} = S_{\gamma^{(p+1)}}$. As before $S_{B(\gamma^{(p)})}$ and $S_{\gamma^{(p)}}$ are both zero. The matrix $A(\gamma^{(p)})$ is given by

$$< R^G(d\gamma^{(p)}(e_i), d\gamma^{(p)}) d\gamma^{(p)} e_i \, , d\gamma^{(p)} > \, ,$$

where e_i, $i = 1,2$ is an orthonormal basis for $T_x M$. Since $\gamma^{(p)}$ is homothetic, this is equal to μg , where $\mu : M \to \mathbb{R}$ depends on

(a) the constant of homothety

(b) the sectional curvature $\text{Riem}^G(d\gamma^{(p)}(e_1), d\gamma^{(p)}(e_2))$ in the Grassmannian, which varies between 0 and 1.

Hence $S_{A(\gamma^{(p)})} = 0$ which implies $S_{\gamma^{(p+1)}} = 0$. Q.E.D.

Suppose now that ϕ maps M harmonically into S^{N-1} so that

$$\Delta\phi = 2e_{\phi}.\phi \ .$$

Suppose further that M is an Einstein manifold, that is M is endowed with a metric of constant curvature. By a theorem of Muto (5) $\gamma^{(1)}$ is homothetic. Therefore, since $\gamma^{(1)}$ is harmonic, by proposition 2.11, $\gamma^{(2)}$ is conformal. In particular

$$[\gamma^{(2)*} {}_k\mathbb{C}^{(2)}]^{(2,0)} = 0 \ .$$

Remark 2.12. The 2'nd Gauss map $\gamma^{(2)}$ will rarely be harmonic. In fact a necessary condition for this is that M be flat (see (1) for details).

REFERENCES

(1) P. Baird and J. Eells, "A conservation law for harmonic maps", Springer Lecture Notes in Math. to appear.

(2) A. Einstein, "Die Grundlage der allgemeinen Relativitätstheorie" Annalen der Physik, 49, 1916.

(3) P. Griffiths and J. Harris, "Algebraic geometry and local differential geometry", Ann. Scient. E.N.S., 4e série, t.12, 1979, p. 355.

(4) D. Hilbert, "Die Grundlagen der Physik", Nachr. Ges. Wiss. Göttingen, (1915), p. 395; & 1917, p. 53.

(5) Y. Muto, "Submanifolds of a Euclidean space with homothetic Gauss map". J. Math. Soc. Japan, vol. 32, N°3 (1980).

(6) M. Obata, "The Gauss map of immersions of Riemannian manifolds in spaces of constant curvature", J. Diff. Geo., 2(1968), p. 217.

(7) E.A. Ruh and J. Vilms, "The tension field of the Gauss map", Trans. A.M.S., 149 (1970), p. 569.

STABILITY AND SELF-DUALITY OF YANG-MILLS FIELDS

J.-P. Bourguignon

Ecole Polytechnique, Palaiseau, France

This talk is a presentation of Part III of (2) which is joint work with H.B. Lawson and to which we refer for any detail. It can be thought of as an illustration of the lecture by J. Eells (this volume).

It deals with Yang-Mills theory for a compact Lie group G on a compact Riemannian manifold M. On the functional space of G-connections over a G-vector bundle over M, one defines the Yang-Mills functional, i.e., the L^2-norm of the curvature.

The functional is weakly compact only when the base space is 2- or 3-dimensional (this result is due to K. Uhlenbeck (3)). The case of 4 dimensions is the most interesting both for physical and for mathematical reasons. From an analytical point of view, we are then exactly in the critical Sobolev range. From an algebraic point of view one has then the special notion of self-dual (or anti self-dual) connections as soon as M is oriented.

In the study of the Yang-Mills functional the main unsolved question is to give a full description of its critical points. They are expected to be absolute minima for "nice" base spaces like the standard sphere S^4. We answer this question in a fairly special case, namely, we prove that local minima are absolute

M. Cahen et al. (eds.), Differential Geometry and Mathematical Physics, 105–106.

minima as soon as M is Riemannian homogeneous and the group G small enough (SU_2, U_2, SO_4, or SU_3).

Our techniques are differential geometric in nature and rely on appropriate Bochner-Weitzenböck formulas which we carefully keep vector-valued.

In reference (1), among other results, we treat the case $M = S^4$ and $G = SU_2$.

REFERENCES

(1) J.-P. Bourguignon, H.B. Lawson, Stability and gap phenomena for Yang-Mills fields, Comm. in Math. Phys. (to appear in 1981).

(2) J.-P. Bourguignon, H.B. Lawson, Yang-Mills theory: its physical origins and differential geometric aspects, (to appear in a volume of the Ann. of Math. Studies edited by S.T. Yau).

(3) K. Uhlenbeck, Connections with L^p-bounds on curvature, (to appear in a volume of the Ann. of Math. Studies edited by S.T. Yau).

EMBEDDED 2-SPHERES IN $\mathbb{R}^3$

H. Chaltin

Université de l'Etat, Mons, Belgique

I. ℓ^2-MANIFOLDS

A Fréchet vector space is a complete topological vector space the topology of which may be defined by a countable set of pseudo-norms. A Fréchet manifold M (modelled on E) is a space locally homeomorphic to a Fréchet vector space E (E is called the model of M). (No differentiability structure is imposed on M).

Any two separable (= having a countable dense subset) infinite-dimensional Fréchet vector spaces are homeomorphic (of course not linearly homeomorphic) ((10)). Hence all such spaces are homeomorphic to one particular among them - considered to be the "standard" Fréchet vector space - viz. ℓ^2. For this reason, a Fréchet-manifold with a separable infinite-dimensional model is simply called an ℓ^2-manifold.

An important property of ℓ^2-manifolds is the following : two ℓ^2-manifolds have the same weak homotopy type iff they are homeomorphic. In particular if a connected ℓ^2-manifold M is such that $\forall\, i \geqslant 1 \quad \pi_i(M) = 0$, then it is homeomorphic to ℓ^2.

The following are classical ℓ^2-manifolds (where M and N are finite dimensional differentiable manifolds, and M compact) :

1) $F(M,N)$, the space of smooth functions $M \to N$ (with the

Whitney C^∞-topology);

(Observe that if M is non-compact, $\mathcal{F}(M,N)$ (always with the Whitney C^∞-topology) is neither separable nor locally homeomorphic to a topological vector space.)

2) $\mathcal{Imm}(M,N)$, the open subset of $\mathcal{F}(M,N)$ of immersions $M \to N$;

3) $\mathcal{E}(M,N)$, the open subset of $\mathcal{Imm}(M,N)$ of embeddings $M \to N$;

4) $\mathcal{D}(M)$, the open subset of $\mathcal{E}(M,N)$ of diffeomorphisms of M (if $\partial M = \phi$, $\mathcal{D}(M) = \mathcal{E}(M,N)$).

There is a natural free action to the right of $\mathcal{D}(M)$ on $\mathcal{E}(M,N)$ (i.e. $(f,g) \to f \circ g$). Denote by $\mathcal{I}(M,N)$ the space of orbits of this action. We may identify the elements of $\mathcal{I}(M,N)$ with the (embedded) submanifolds of N, diffeomorphic to M. We shall denote by im the identification map $\mathcal{E}(M,N) \to \mathcal{I}(M,N)$. Again, $\mathcal{I}(M,N)$ is an ℓ^2-manifold, modelled on the space of sections of the normal bundle in N of a particular copy of M in N. (Observe that the ℓ^2-manifolds $\mathcal{D}(M)$, $\mathcal{E}(M,N)$ and $\mathcal{I}(M,N)$ have (formally) different models).

The triple $(\mathcal{E}(M,N), \text{im}, \mathcal{I}(M,N))$ is a locally trivial principal $\mathcal{D}(M)$-bundle ((2),(21)) (hence e.g. $\varinjlim_n \mathcal{I}(M,\mathbb{R}^n)$ is a classifying space for $\mathcal{D}(M)$ - (21)).

Although there seems to be no satisfactory way to put a differentiability structure on these spaces, the following is a partial substitute (cfr. (21)). Let S be a finite dimensional differentiable manifold. A map $f : S \to \mathcal{F}(M,N)$ is smooth iff the evaluation map $S \times M \to N : (s,m) \to (f(s))(m)$ is smooth also. A map $f : S \to \mathcal{I}(M,N)$ is smooth iff it has smooth liftings locally (i.e. $\forall x \in S \ \ \exists U \ \ \exists g : U \to \mathcal{E}(M,N)$ s.t. $\text{im} \circ g = f_{|U}$).

An element $S \in \mathcal{I}(S^n,N)$ is called an "(embedded) n-sphere in N".

II. SMALE'S CONJECTURE

As any linear map $A \in O(n+1)$ leaves S^n invariant, we may identify $O(n+1)$ with a subgroup of $\mathcal{D}(S^n)$.

Proposition 1. $O(2)$ *is a strong deformation retract of* $\mathcal{D}(S^1)$.

Proof. Define $G = \{\phi \in \mathcal{D}(S^1) | \phi(0) = 0$ and $\phi'(0) > 0\}$.

Clearly $\mathcal{D}(S^1) = O(2).G$ (group product) and $O(2) \cap G = \{id_{S^1}\}$.

Hence $\mathcal{D}(S^1) \cong O(2) \times G$ (cartesian product). But clearly G is contractible by the linear retraction $G \times I \to G$:

$(\phi,t) \to t\phi + (1-t)id$.■

Remark. We have in fact used that $S^1 \cong SO(2)$ and that $(GL(1,\mathbb{R}))°$ is convex. This situation, of course, is peculiar to dimension 1!

Theorem 2. (Smale (24)). $O(3)$ *is a strong deformation retract of* $\mathcal{D}(S^2)$.

The following has been known as "Smale's conjecture" (cfr. (2)):

Conjecture 3. $O(4)$ *is a strong deformation retract of* $\mathcal{D}(S^3)$.

The following statements are equivalent to this conjecture:

Conjecture 4. $\forall i \geq 1$, $\pi_i(I(S^2,\mathbb{R}^3),S^2) = 0$ (cfr. (3),(8)).

Conjecture 5. $\forall i \geq 1$, $\pi_i(I(S^2,\mathbb{R}^3 \setminus \{0\}),S^2) = 0$ (cfr. (4),§8).

Remark . Cerf ((4)) proved that

1) $\pi_o(\mathcal{D}(S^n)_{/O(n+1)}) = \Gamma^{n+1}$ for $n \geq 5$;

2) $\pi_1(\mathcal{D}(S^n)_{/O(n+1)})$ is an extension of Γ^{n+2} for $n \geq 5$

(recall that $\Gamma^n = 0$ if $n \leq 6$, $\Gamma^7 = \mathbb{Z}_{28}$, ...). Hence the analogue of Smale's conjecture in dimensions ≥ 5 is false (One might, however, conjecture that $\mathcal{D}(S^4)_{/O(5)} \simeq$ ☆ ("Smale + 1")).

Theorem 6. (Alexander (1), Morse and Baiada (20), Cerf (3)) (The "differentiable Schönflies conjecture in dimension 3"). $I(S^2,\mathbb{R}^3)$ *is connected (equivalently, any* $S \in I(S^2,\mathbb{R}^3)$ *bounds a 3-disk).*

Remarks. 1) Schönflies proved that any simple closed curve in $\mathbb{R}^2$ bounds a 2-disk and conjectured that any subset of $\mathbb{R}^3$ homeomorphic to S^2 bounds a 3-disk.

2) Alexander found a counter-example to this conjecture (the "Alexander horned sphere") and proved the conjecture for PL-spheres.

Theorem 7. (Cerf (3)). $\pi_1(I(S^2,\mathbb{R}^3),S^2) = 0$.

(Cerf constructs a section $\sigma : I(S^2,\mathbb{R}^3) \to I(S^2,\mathbb{R}^3)^{\sim}$.)

Corollary. $\Gamma^4 = 0$.

The importance of this corollary lies, among other things, in the fact that it was the "missing link" in the proof of

Theorem 8. (Tom, Munkers, Milnor, Kervaire, Smale, Cerf) (cfr. (12), Kuiper in (3) (p. IX)). *Any PL-manifold of dimension* $n \leqslant 7$ *admits a differentiability structure, unique if* $n \leqslant 6$.

Remark. These bounds were already known to be sharp, because of the existence of an 8-dimensional compact PL-manifold which is not smoothable ((17), p. 248) and because of the existence of "exotic" 7-spheres ((15),(11)).

If one replaces "embedded" by "immersed", theorem 7 (and hence statement 4) is false. Indeed Smale (again !) proved

Theorem 9. ((23)). *$Imm(S^2,\mathbb{R}^3)$ is connected (" a sphere may be turned inside out through immersions".*

Cfr. (22),(18),(19)).

Corollary. *The space of "immersed 2-spheres in $\mathbb{R}^3$". $Imm(S^2,\mathbb{R}^3)_{/\mathcal{D}(S^2)}$ has a connected 2-fold covering $Imm(S^2,\mathbb{R}^3)_{/\mathcal{D}^+(S^2)}$ (the space of "oriented immersed 2-spheres in $\mathbb{R}^3$ ") (here $\mathcal{D}^+(S^2) = \{\phi \in \mathcal{D}(S^2) \mid \phi$ is orientation preserving$\}$).*

Hence any proof of Smale's conjecture must rely on the fact that the spheres under consideration are embedded.

III. THE SPACE $I = I(S^2,\mathbb{R}^3)$

Let us call any element of I a "sphere".

Define $h : \mathbb{R}^3 \to \mathbb{R} : (x,y,z) \to z$ (the "height function")

$h_{\star} : I \to F(S,\mathbb{R}) : S \to h_{|S}$

Σ_S = the set of singular points of $h_{|S}$.

The "natural stratification" ((3),(4)) on each of the spaces $F(S,\mathbb{R})$ $(S \in I)$ induces, by means of the map $h_{\star}$, a stratification $(I_i)_{i\in\mathbb{N}}$ on I.

(Recall that $(I_i)_i$ has the following properties (cfr. also (3), (4) and (5)):

1) $S \in \bigcup_{i \in \mathbb{N}} I_i \Rightarrow \# \Sigma_S < \aleph_o$ (in particular $I_\infty = I \setminus (\bigcup_{i \in \mathbb{N}} I_i) \neq \phi$);

2) I_i is open and dense in $(\bigcup_{i \geq n} I_i) \cup I_\infty$;

3) $I_{i+1} \subset \overline{I}_i \setminus I_i$;

4) I_i is a codimension i submanifold of I (in particular $I \setminus I_\infty$ is a deformation retract of I).)

We shall describe only I_o and $I_1 = I_{1,\alpha} \cup I_{1,\beta}$ (cfr. also (3) and (4)):

$S \in I_o$ iff 1) $\forall p \in \Sigma_S$ p is a Morse-singularity;

2) $h_{|\Sigma_S}$ is 1-1.

(A sphere $S \in I_o$ is called "generic" - cfr. (6)).

$S \in I_{1,\alpha}$ iff 1) $\exists!\, q \in \Sigma_S$ such that q is a singularity of type $x^2 + y^3$;

2) $\forall p \in \Sigma_S \setminus \{q\}$ p is a Morse-singularity;

3) $h_{|\Sigma_S}$ is 1-1.

(Cfr. fig. 1, middle sphere).

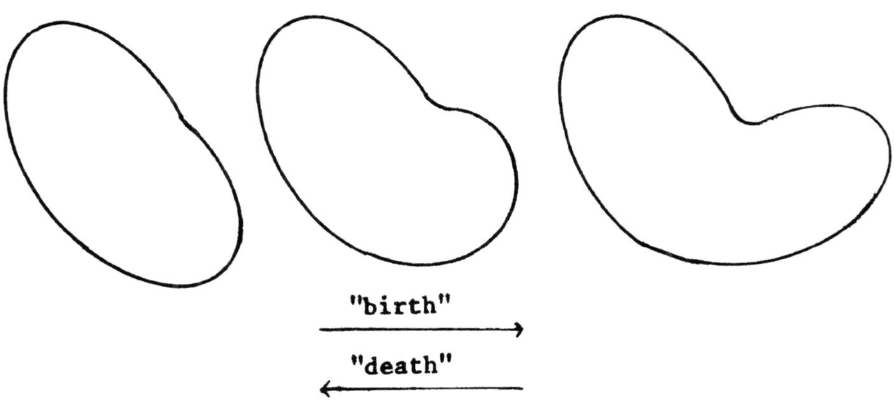

fig. 1

$S \in I_{1,\beta}$ iff 1) $\forall p \in \Sigma_S$ p is a Morse-singularity;

2) $\# h(\Sigma_S) = \# \Sigma_S - 1$ (i.e. exactly two points $p,q \in \Sigma_S$ have same h-value).

(Cfr. fig. 2, middle sphere).

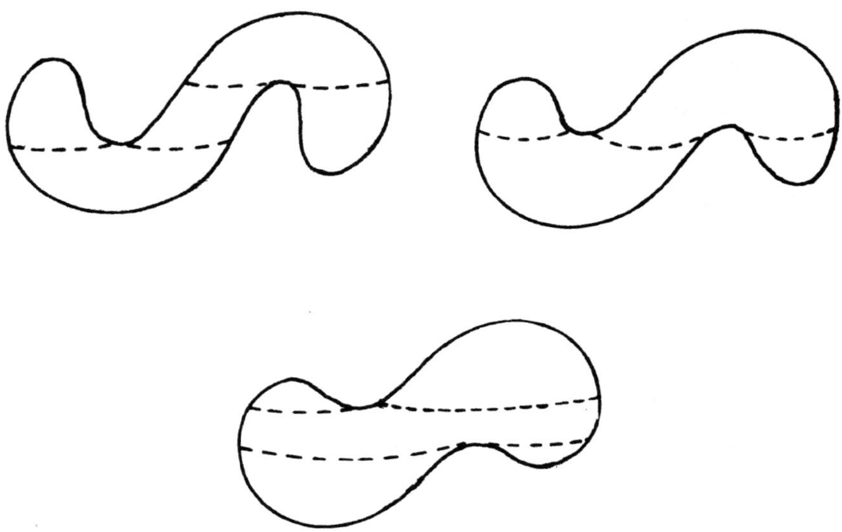

fig.2

A sphere $S \in I_1$ corresponds to a ("generic") transition between two different connected components of I_o (fig. 1 and 2). The transition corresponding to a sphere $S \in I_{1,\alpha}$ is called a "birth" or a "death" (according to the direction in which $I_{1,\alpha}$ is crossed).

<u>Theorem</u> 10. (Morse; cfr. (16)). *Let* M(S) *(resp.* m(S) *, resp.* s(S)) *denote the number of maxima (resp. minima, resp. saddle-points) on* $S \in I_o$. *Then* $M(S)-s(M)+m(S) = \chi(S) = 2$.

<u>Corollary</u>. $\# \Sigma_S = 2s(S) + 2$.

s(S) "measures", in a rather rough way, the "complexity" of the sphere S. We shall call s(S) the "Alexander complexity" of S (because Alexander's proof ((1)) proceeds by induction on s(S) - or rather the PL-equivalent of s(S)).

If s(S) = 0 (in which case we shall say S is "almost standard") S lies in the same connected component as the standard sphere S^2.

If $s(S) = 1$, then S can have two different types (corresponding to two distinct components of I_o) (fig. 3-from now on, most spheres will be drawn with the (symmetric) front half removed).

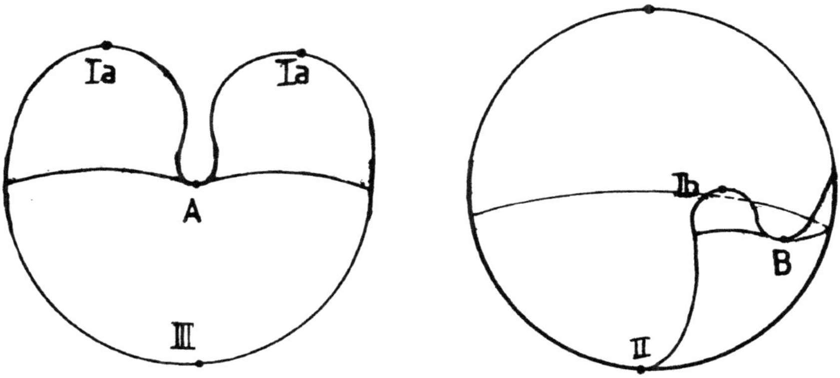

fig. 3

Classification of Morse-singularities

1) If p is saddle-point on $S \in I_o$, then the connected component of $h^{-1}(h(p))$ containing p (denoted $\Gamma(p)$) can have two different froms (looked upon from above) :

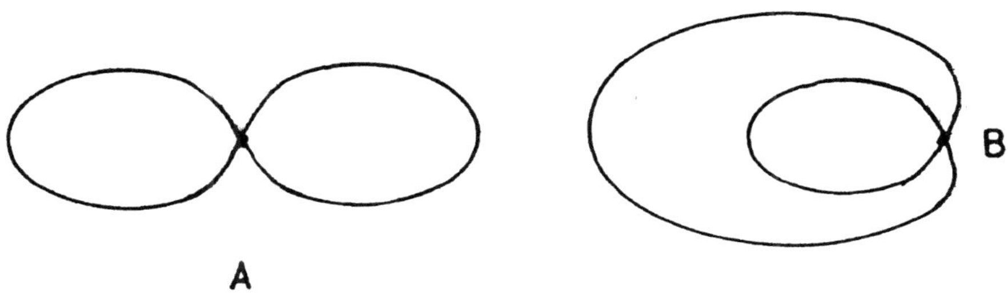

fig. 4

2) Let p be a maximum (the discussion is similar for a minimum, of course). Denote by Δ_p the connected component of $S \setminus \bigcup_{p \in \Sigma_S} \Gamma(p)$ contiguous to p. There is exactly one other $q \in \Sigma_S$ in $\partial\Delta_p$. If q is a minimum, then $s(S) = 0$, i.e. S is almost standard. Suppose then that q is a saddle-point. Then a neighbourhood of $\Delta_p \cup \Gamma(q)$ is of one of the five different types shown in fig. 5

below.

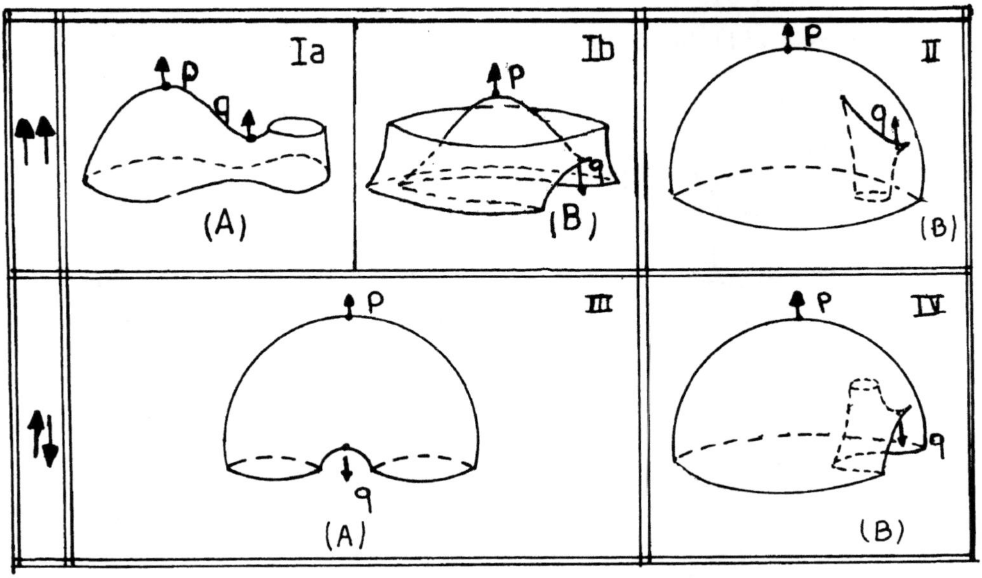

fig. 5

There is a natural dichotomy between types I and II (same orientation of the normals in p and g) and types III and IV (opposite orientation). The dichotomy defined by the type (A or B) of the saddle-point q is less satisfactory for our purposes, however, as it would distinguish between types Ia and Ib, which we want to keep together (it will be clear why in the next section)

All five types occur in the s = 1 spheres of fig. 3.

IV. SIMPLIFICATIONS

If a sphere $S \in I_o$ has a maximum (or minimum) p of type I (Ia or Ib), there is a - geometrically easily conceivable- homotopy in $I_o \cup I_{1,\alpha}$ starting from S, passing through a "death" $\in I_{1,\alpha}$ (where p and q "die", or "annihilate each other") and ending at a sphere S' with s(S') = s(S)-1 (one simply pushes down the "bump" Δ_p - fig. 6).

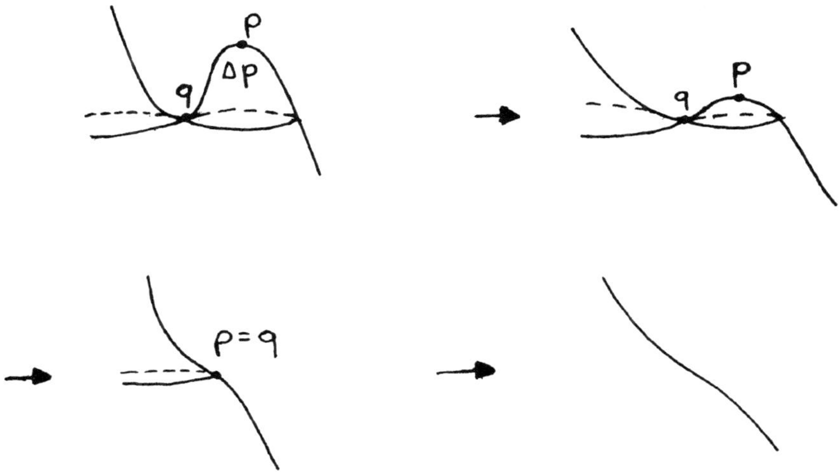

fig. 6

Such a homotopy will be called an "inessential simplification". (An extremum of another type does not allow such a construction - the types III and IV having an obvious obstruction : the different orientation of the normals at p and g).

A sphere, which can be reduced to an s = 0 sphere through a succession of inessential simplifications will be called "reducible".

Unfortunately, many spheres do not admit any inessential simplifications at all, e.g. the sphere drawn in fig. 7.

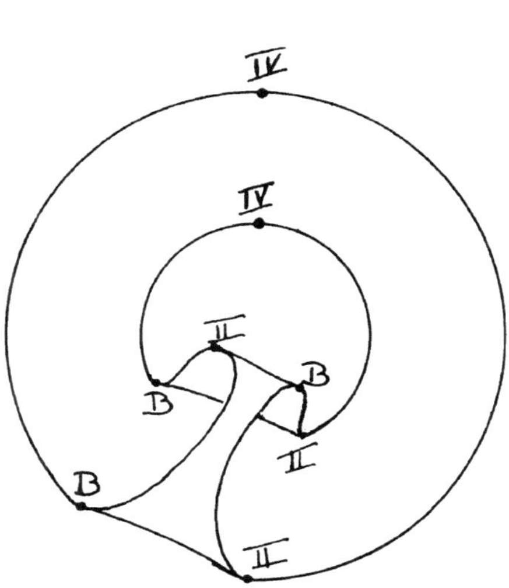

fig. 7

(the "embryo", alias the "Chinese mushroom") (In some sense, this is the simplest possible counterexample).

This means that if one wants to develop a strategy of simplifications in order to reduce any sphere $S \in I_o$ to the standard sphere, taking only inessential simplifications (or "deaths") will not suffice. Observe that although the Alexander complexity of a sphere is reduced by 1 when an inessential simplification is applied, it is "insensitive" to the $I_{1,\beta}$-transitions. This means that the Alexander complexity will not serve our purposes, as will be explained below.

Our aim is to develop a strategy of "essential" simplifications (i.e. certain transitions of type $I_{1,\beta}$) which will reduce any sphere to the standard sphere. In order to achieve this, we shall have to decide which (type $I_{1,\beta}$) transitions will have the title of "simplification" (e.g. of the two possible orientations of a transition only one can be designated a "simplification" - the other one becoming automatically a "complication").

It will be convenient to define a new "complexity" $\alpha : I_o \to \mathbb{N}$ adapted to this strategy in the sense that

1) applying any essential simplification to S will reduce $\alpha(S)$ by at least 1;

2) if $\alpha(S) = 0$, S "more or less looks like a standard sphere", i.e. S is reducible.

V. TUBES AND DIGRAPHS

After some inspection, it is clear that type B saddle-points make more complicated spheres than type A saddle-points do. Indeed, suppose $S \in I_o$ has only type A saddle-points. Then S is "horizontally convex" in the following sense : let D be the 3-disk bounded by S. Then $H \cap \overset{\circ}{D}$ is a union of open disks for any horizontal plane H. Of course, this is not true anymore as soon as there is a type B saddle-point.

Moreover, if $S \in I_o$ has only type A saddle-points, it is easy to see that S is always reducible. On the other hand, as we already have noticed (cfr. fig. 7), it is possible to construct irreducible spheres with only type B saddle-points.

Now if one looks at the sphere S to the left of fig. 3, all points of S seem to lie "on the outside" of S. On the contrary, one has the intuitive feeling that the sphere on the right has two parts : an "outside" and an "inside" one (cfr. fig. 8). But where exactly should we put the boundary between the two parts? For technical reasons, the best choice for this boundary seemed to us to be the "inner loop" r_p of the figure-eight level-manifold $\Gamma(p)$ (cfr. fig. 8) (although this choice is perhaps not the most satisfactory one for the eye).

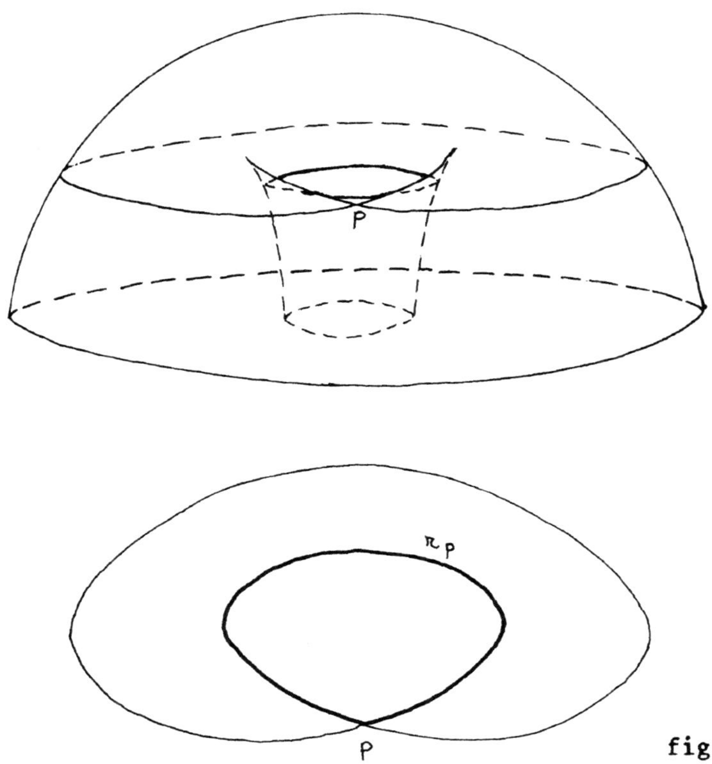

fig.8

One is naturally tempted to define a new complexity on I_o by $S_B(S)$ = number of type B saddle-points on S. It turns out that this choice would not be very appropriate. Indeed the spheres in fig. 7 and fig. 9 e. g. both have a number of type B saddle-points equal to 3. With the above definition both spheres would have equal complexity. Yet the sphere of fig. 9 is reducible and, as we already explained, we want to consider reducible spheres as "simpler" spheres than the non-reducible ones.

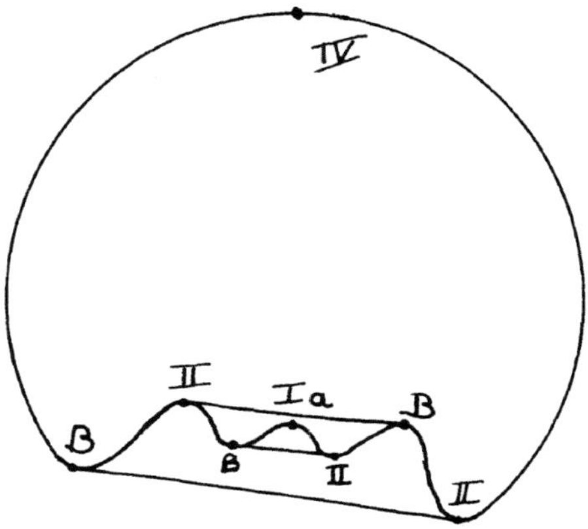

fig. 9

Instead, we shall have to proceed to a more careful study of how the type B saddle-points affect the geometry of a sphere.

Definitions. A "rim" on S is a (PD-) simple closed curve of the form r_p (cfr. fig. 8) for some type B saddle-point p.

Denote by R_S the union of rims on S.

A "tube" on S is a connected component of $S \setminus R_S$.

If T_1 and T_2 are tubes, define $T_1 \ll_S T_2$ (short : $T_1 \ll T_2$) iff there is a (by proposition 11 unique) rim $r \subset \overline{T_1} \cap \overline{T_2}$ such that T_2 lies "inside r" and T_1 "outside r" (cfr. fig. 10)

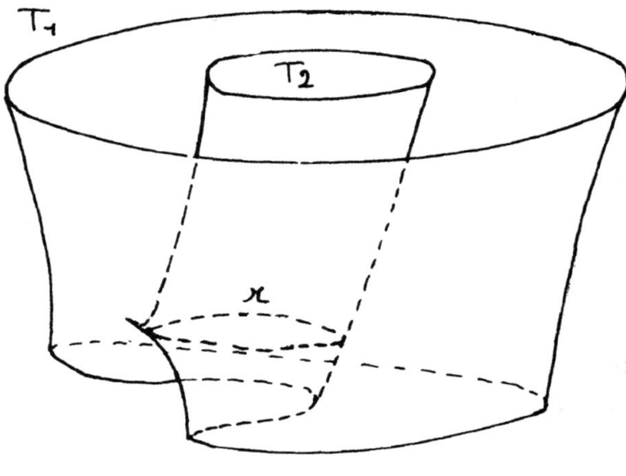

fig. 10

Denote by $\tau(S)$ the digraph (= directed graph) $(\pi_o(S \setminus R_S), \preccurlyeq_S)$ and by $T(S)$ the isomorphism class of digraphs determined by $\tau(S)$.

<u>Proposition</u> 11. $\tau(S)$ *is a (finite)* 1-*connected digraph (a "directed tree").*

<u>Proof</u>. (Essentially the Jordan curve theorem). Denote by G the (countable) set of isomorphism classes of finite 1-connected digraphs. We shall consider T as a G-valued complexity on I_o. (<u>Remark</u> : the map T is not onto). Observe that $s_B(S) = \# T(S)-1$, i.e. we may recover $s_B(S)$ from $T(S)$, which contains much more (and enough) information about S than $s_B(S)$.

We shall regard a sphere as being built up by means of its constituent tubes (i.o.w. the tubes of a spheres are its elementary building blocks); cfr. e.g. fig. 11 and 12 for the sphere of fig.7

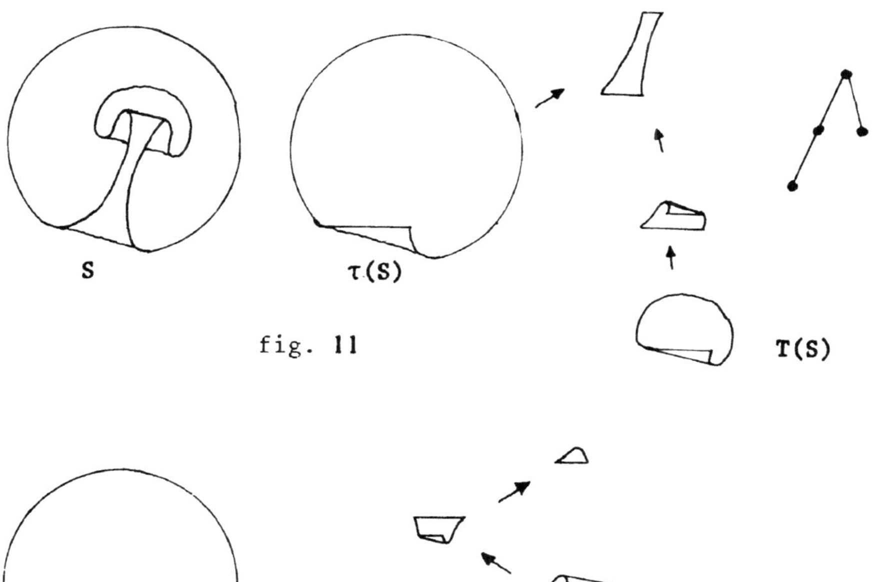

fig. 11

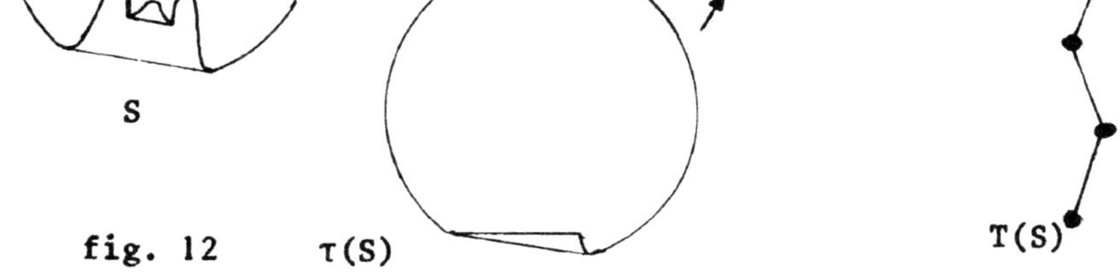

fig. 12

and 10 respectively, and fig. 13 through 15 for some other examples (Observe that ≪-minimal tubes correspond to roughly "spherical" parts).

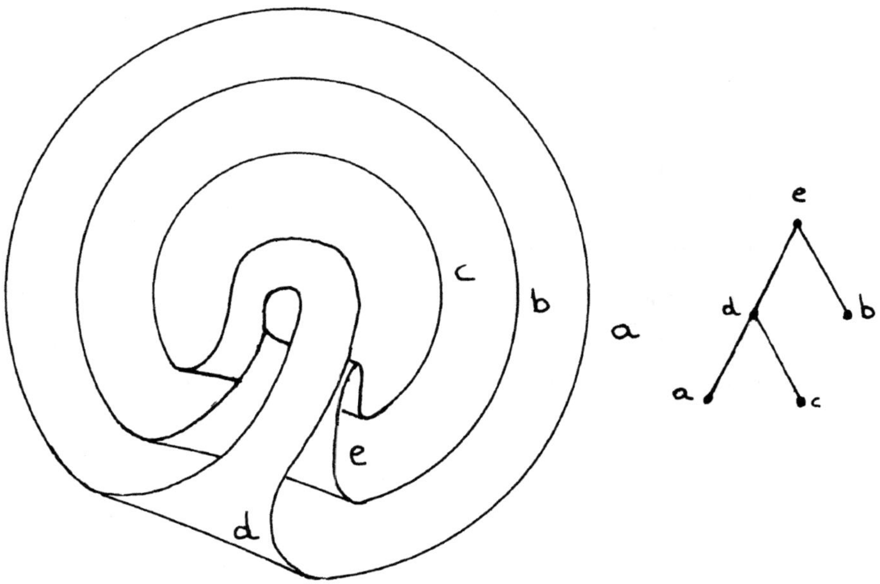

fig. 13

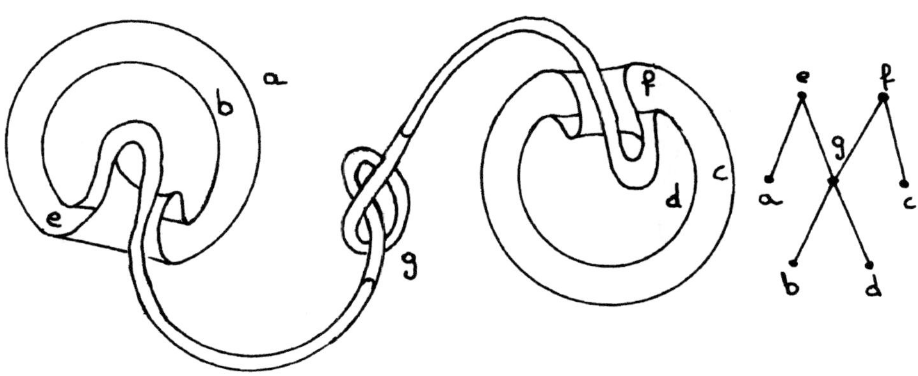

fig. 14

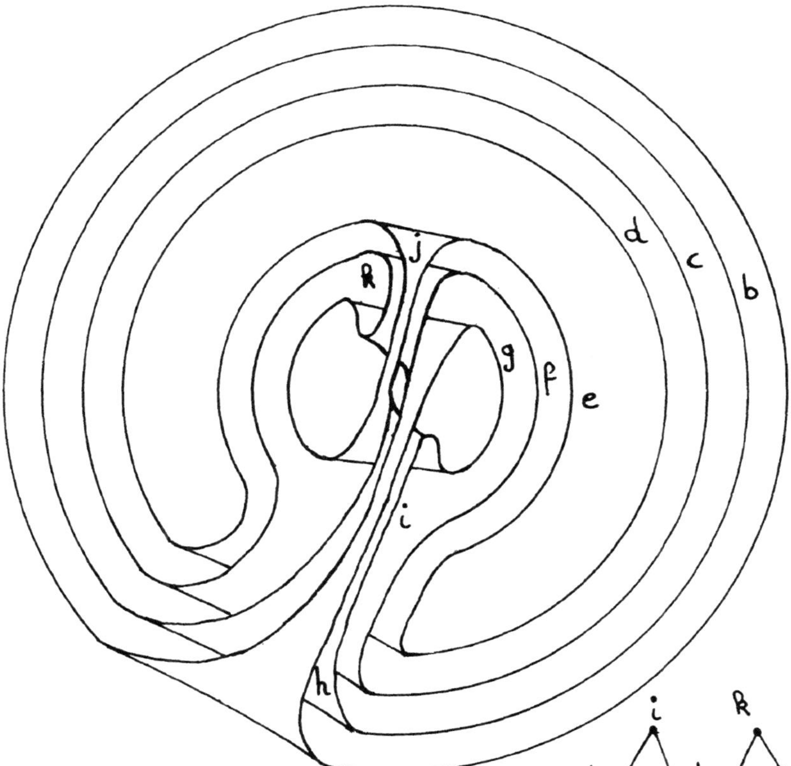

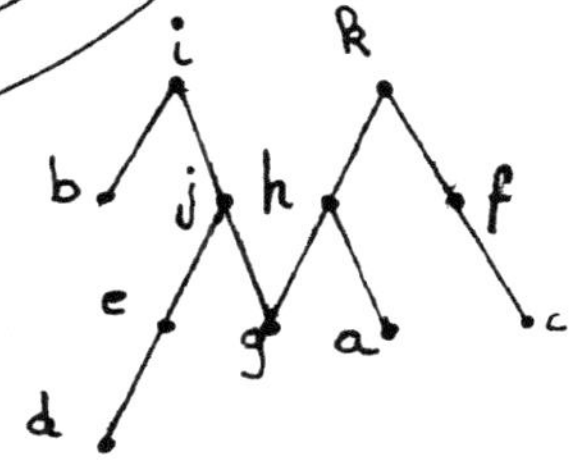

fig. 15

Our next step will be to translate the transitions (corresponding to connected components of I_1) between connected components of I_o into relations on G.

On finite directed trees, define the following operations (where $a,b,c \in X$) :

1) $K_a(X,R) = (X \setminus \{a\}, R_{|X \setminus \{a\}})$ if a is terminal in (X,R) (i.e. $\exists !\, b \;\; b\, R\, a, \;\; \nexists\, c \;\; a\, R\, c$) (cfr. fig. 16);

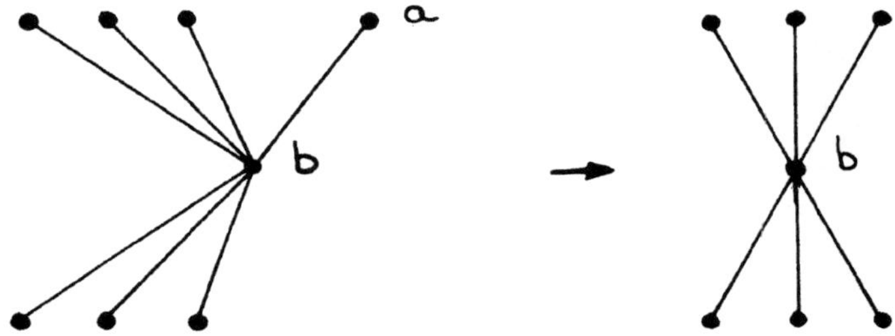

fig. 16

2) $L^{\star}_{a,b}(X,R) = (X_{/a=b}\ , R_{/a=b})$ if c (which is then unique) such that a R^{-1} c R b (cfr. fig. 17);

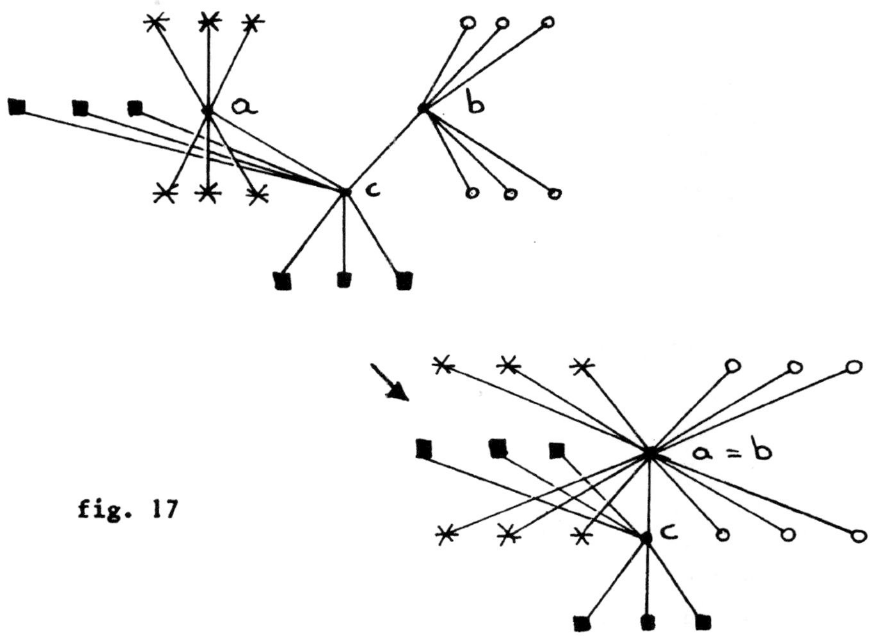

fig. 17

3) $I_{a,b,c}(X,R) = (X_{/a=c}\ , (R \setminus \{(b,c)\})_{/a=c})$ if a R b R c (cfr. fig. 18).

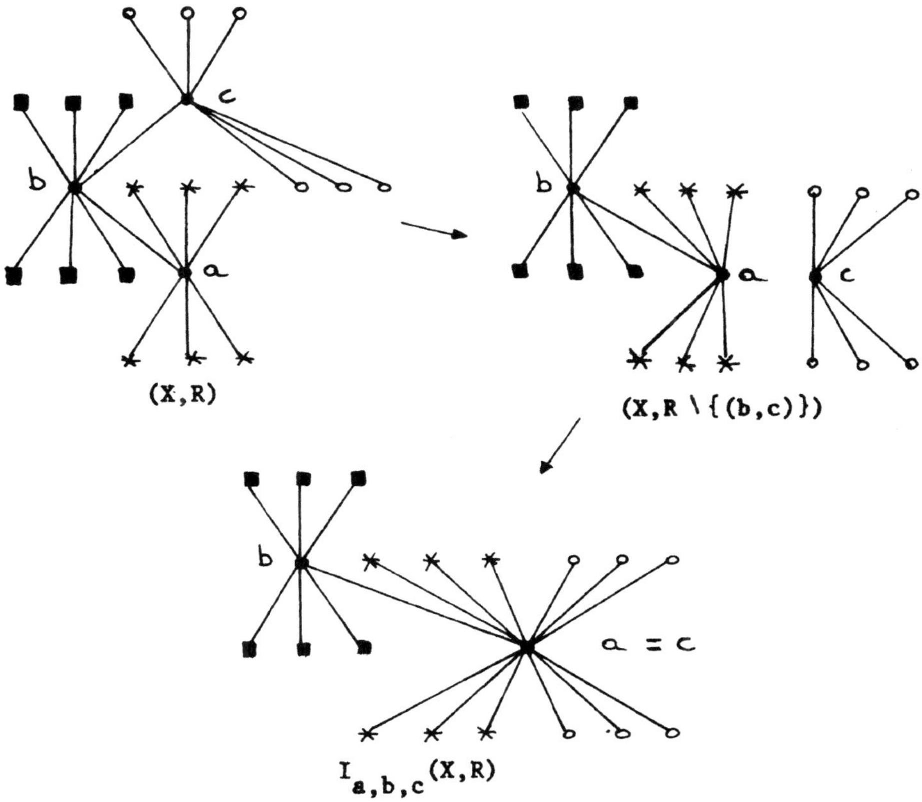

fig. 18

Next "lift" these operations as relations K, L^{-1} and I on G. E.g. if Ξ and $\Upsilon \in G$, $\Upsilon \; I \; \Xi$ iff $\exists (X,R) \in \Xi \; \exists a,b,c \in X$ such that $I_{a,b,c}(X,R) \in \Upsilon$. (Observe that $\Xi \; L \; \Upsilon$ iff $\exists (X,R) \in \Xi \; \exists a,b \in X$ s.t. $L^{☆}_{a,b}(X,R) \in \Upsilon$).

If $(\Xi,\Upsilon) \in K \cup L^{-1} \cup I$, then $\# \Xi = \# \Upsilon - 1$; hence $(K \cup L^{-1} \cup I)\hat{}$ (the transitive-and-reflexive closure of $K \cup L^{-1} \cup I$) is an order relation on G.

Proposition 12. *Let $\sigma \to I_o \cup I_1$ be a path crossing $I_{1,\alpha}$ (resp. $I_{1,\beta}$) once, i.e. $\sigma(I \setminus \{\frac{1}{2}\}) \subset I_o$ and $\sigma(\frac{1}{2}) \in I_{1,\alpha}$ (resp. $I_{1,\beta}$). Then* $(T(\sigma(0)), T(\sigma(1)) \in K \cup K^{-1} \cup id_G$ (resp.$\in I \in I^{-1} \cup L \cup L^{-1} \cup id_G$) *In other words, the relations* I, L, K, I^{-1}, L^{-1}, K^{-1} *and* id_G *"translate" the generic transitions between components of I_o into* *relations on* G.

For examples of transitions of types K, L and I, cfr. fig.19 through 21.

fig. 19

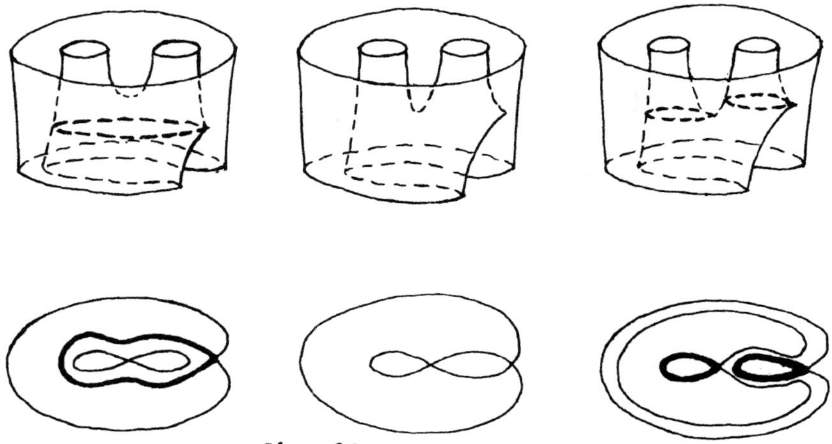

fig. 20

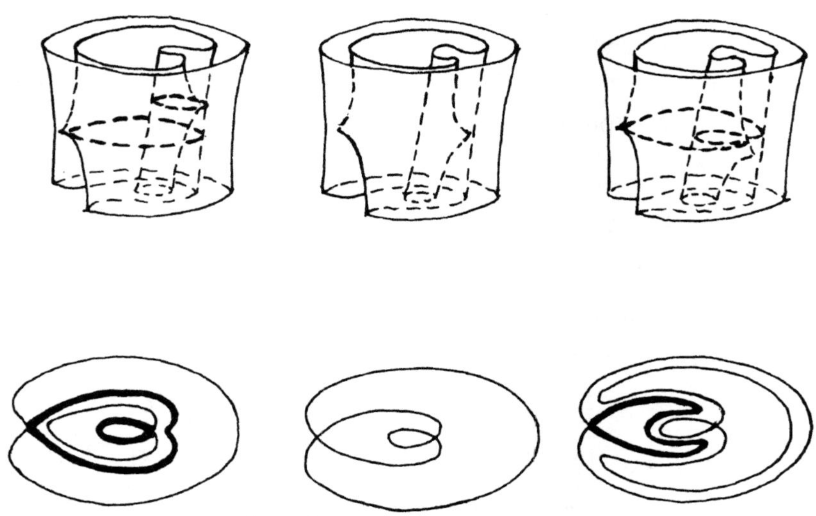

fig. 21

Main proposition 13. $\leqslant = (I \cup (L \setminus K^{-1}) \cup K)^{\wedge}$ *is an order relation on* G *such that each digraph has a finite number of predecessors only. Moreover the one point digraph is the only* $\leqslant$*-minimal element in* $T(I_o)$. *(There are infinitely many* $\leqslant$*-minimal elements in* G*).*

Hence the "strategy" : "replace a digraph Ξ in $T(I_o)$ by any digraph $\Upsilon < \Xi$", will automatically yield the one point digraph after a finite number of steps. If we could lift this strategy to geometry (which we can), this would yield the strategy we were looking for in section 4. This is precisely the reason we have to dismiss $(I \cup L^{-1} \cup K)^{\wedge}$ (which is an order-relation for trivial reasons) as a "strategy" (it does not lift to geometry).

Define at last $c(\Xi) = \max \{n \mid \exists\, \Xi_o < \ldots < \Xi_n = \Xi\}$.
We may then define an $\mathbb{N}$-valued (instead of G-valued) complexity α on I_o by $\alpha(S) = c(T(S))$. This is the "good" way to extract from $T(S)$ an $\mathbb{N}$-valued complexity ($s_B(S) = \# T(S) - 1$ was the "bad" way).

VI. A QUOTIENT SET OF G

Proposition 14. *Any* $\leqslant$*-terminal tube in* S *may be removed by a succession of inessential simplifications. In other words, a* K*-operation is a composition of inessential simplifications* (cfr. fig. 22).

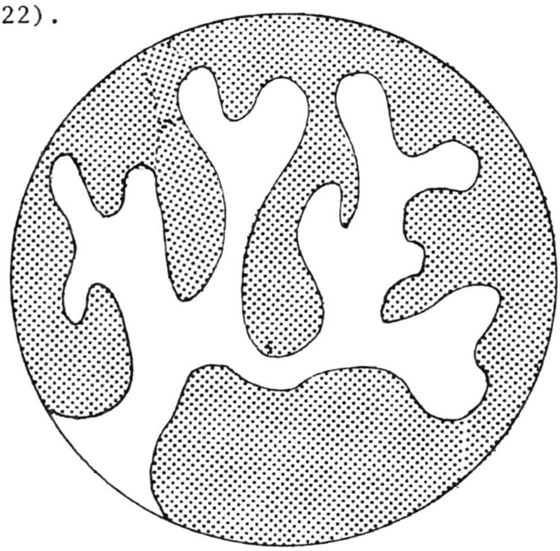

fig. 22

This suggests that if we disregard inessential simplifications, we should also disregard K-operations, i.e. we should quotient G over $(K \cup K^{-1})^{\wedge}$, the equivalence relation generated by K.

For technical reasons, it seemed convenient to disregard L-operations also, although they usually involve type $I_{1,\beta}$ transitions. Indeed, they do not affect too seriously the geometry of a sphere.

More precisely, let $R = (K \cup L \cup K^{-1} \cup L^{-1})^{\wedge}$ be the equivalence relation generated by $K \cup L$ and let $\pi : G \to G_{/R} = \Gamma$ be the identification map. If T is any relation on G, denote by $\pi_{\star}(T)$ the relation induced by π on Γ.

<u>Proposition</u> 15. *There is a partition* $I_o \cup I_1$ *of* I *such that*

1) $\pi_{\star}(I_o) = id$;

2) $\leqslant\ = (\pi_{\star}(I_1))^{\wedge}$ *is an order relation on* Γ *such that any element has a finite number of predecessors only and such that* $\pi(T(I_o))$ *contains only one minimal element.*

(I_1 is the "essential" part of I).

We shall call the quantity $\pi(T(S)) \in \Gamma$ the "essential complexity" of $S \in I_o$. As in section 5, it is again possible to associate to this Γ-valued complexity an $\mathbb{N}$-valued one, viz. $\beta(S) = c(\pi(T(S)))$, where c is defined on Γ in an analoguous way.

VII. CONCLUSION

Our aim is to extend our machinery (which was here defined for spheres $S \in I_o$ only) to spheres in $I \setminus I_{\infty}$(recall that $I \setminus I_{\infty}$ was a deformation retract of I) and to describe a deformation retraction of I^{n+1} into I^n (where $I^n = \{S \in I \setminus I_{\infty} \mid \beta(S) \leqslant n\}$. It is not hard to see that I^o (the space of reducible spheres) is contractible (indeed only inessential simplifications need to be applied, and this can be done in a canonical way).

VIII. APPENDIX

Our methods apply to other spaces of embedded submanifolds as well. We give three examples.

Let $M \subset V \subset W$ (dim M = dim V = 2, dim W = 3, M compact and possibly with boundary).

(I) If W is a line bundle over V, define

$\mathcal{A} = \{N \in \mathcal{I}(M,W) \mid N$ is 1-tangent to M along $\partial N = \partial M\}$

$\mathcal{A}^o$ = the connected component of $\mathcal{A}$ containing M.

Then $\mathcal{A}^o$ is contractible. (In the proof the norm function $W \to \mathbb{R}$ defined by a particular Riemannian (bundle) metric on W plays a role analogous to that of the height function $h : \mathbb{R}^3 \to \mathbb{R}$ of the previous sections).

(II) If W is an M-bundle over S^1 and $M = V$ a particular fiber of W, put

$\mathcal{B} = \{N \in \mathcal{I}(M,W) \mid N \cap \partial W = \partial N\}$

$\mathcal{B}^o$ = the connected component of $\mathcal{B}$ containing M.

Then $\mathcal{B}^o$ deformation retracts onto $\{N \mid N$ is a fiber of $W\} \cong S^1$. (Here the bundle projection plays a role analogous to that of $h : \mathbb{R}^3 \to \mathbb{R}$).

These facts imply the following results :

1) $\mathcal{D}(M \times I) \simeq \mathbb{Z}_2 \times \mathcal{D}(M)$ ("the space of pseudo-isotopies of M is contractible").

2) $\mathcal{D}(M \times S^1)$ contains an open subgroup $G \simeq S^1 \times \mathcal{D}(M) \times \Omega\mathcal{D}(M)$ (similar results hold if $M \times S^1$ is replaced by an arbitrary M-bundle over S^1).

3) If W is a "solid torus" (i.e. homeomorphic to $M \times I$, with $\partial M \neq \phi$, but possibly non orientable), then any connected component of $\mathcal{D}_\partial(W) = \{\phi \in \mathcal{D}(W) \mid \phi_{|\partial W} = id_{|\partial W}\}$ is contractible.

Remark. 1 (for $M = S^2$) and 3 (for $W = D^3$) are equivalent to Smale's conjecture. Hatcher has proved that 1,3 (for $M \neq P^2$ ((7)) and 2 (for $M \neq S^2$, $M \neq P^2$ ((7)) and for $M = S^2$ ((9))) are implied by Smale's conjecture.

4) For arbitrary W, $\mathcal{I}(D^2 \times S^1, W)$ (which is a union of

connected components of $\mathcal{I}(S^1 \times S^1, W)$) has the same homotopy type as $\{A \in \mathcal{I}(S^1,W) \mid A$ has trivial normal bundle in $W\}$.

5) If W is a line bundle over V ($\partial V = \phi$), $\{A \in \mathcal{I}(S^1,W) \mid A$ unknots in $W\}$ (hence also the corresponding component of $\mathcal{I}(S^1 \times S^1,W)$) has a covering space which has the homotopy type of PTW (the projective tangent bundle of W).

(III) If $M \subset \partial W$, define

$$\mathcal{C} = \{N \in \mathcal{I}(M,W) \mid N \cap \partial W = M\}$$

$$\mathcal{C}^o = \{N \in \mathcal{C} \mid N \text{ is homotpic in } \mathcal{I}(M,W) \text{ to } M\}.$$

Thenagain $\mathcal{C}^o$ is contractible.

The assertion in I remains true if W is a non-trivial S^1-bundle over V, having a section. Also, if M = V and $\chi(M) < 0$, $\mathcal{D}(W)$ has an open subgroup which has the homotopy type of a covering space of $\mathcal{D}(M)$ (cfr. 2).

Similarly, if W is a non-trivial I-bundle over M, $\mathcal{D}(W)$ has the homotopy type of a 2-fold covering of $\mathcal{D}(M)$ (which is trivial if $\chi(M) < 0$) (cfr. 1) and, if $\chi(M) < 0$, each connected component of $\mathcal{D}_\partial(W)$ is contractible (cfr. 3).

In future work, we hope to combine I and III with the techniques developed in the proof alluded to in section 7 in order to extend the results in I to more general situations (e.g. for arbitrary W (not necessarily P^2-irreducible, as in (7)) if $\partial M \neq \phi$). We would also like to compare our methods with Laudenbach's disjuntion technique ((13),(14)) and Hatcher's "slice" technique ((7),(9)).

We would like to thank Françoise Derroitte of the Art School "75" at Woluwe-Saint-Lambert and Isabelle Bony of the Académie Royale des Beaux-Arts de Bruxelles for their patient execution of the drawings in this article.

REFERENCES

(1) J.W. Alexander, "On the subdivision of 3-space by a polyhedron", Proc. Nat. Acad. Sc. U.S.A., t. 10, 1924, p. 6-8.

(2) J. Cerf, "Topologie de certains espaces de plongements", Bull. Soc. Math. France, t. 89, 1961, p. 226-380 (Thèse Sc. Math. Paris, 1960).

(3) J. Cerf, "Sur les difféomorphismes de la sphère de dimension trois ($\Gamma_4 = 0$)", 1968, Lecture Notes in Math. # 53, Springer.

(4) J. Cerf, "La stratification naturelle des espaces de fonctions différentiables réelles et le théorème de la pseudo-isotopie", Publications Mathématiques de l'I.H.E.S. n° 39 p. 5 à 173.

(5) C.G. Gibson, K.W. Wirthmüller, A.A. du Plessis and E.J.N. Looijenga : Topological stability of smooth mappings, 1976, Lecture notes in Math. # 552, Springer.

(6) Martin Golubitsky and Victor Guillemin : Stable mappings and their singularities, 1973, Graduate texts in Math. # 14, Springer.

(7) Allen E. Hatcher : Homeomorphisms of sufficiently large P^2-irreducible 3-manifolds, Topology 15 (1976), 343-347.

(8) Allen E. Hatcher : Linearization in 3-dimensional topology, Proceedings of the International Congress of Mathematicians Helsinki, 1978, pp. 463-468.

(9) Allen E. Hatcher : On the diffeomorphism group of $S^1 \times S^2$, preprint.

(10) David W. Henderson : Infinite dimensional manifolds are open subsets of Hilbert space, Topology Vol. 9, pp. 25-33.

(11) Kervaire M. and Milnor, J.W. : Groups of homotopy spheres : I, Annals of Math. 77 (1963), 504-537.

(12) N.H. Kuiper, Mathematical Review N° 6641 a-d, vol. 33, p. 1126 concerning J. Cerf "La nullité de π_o (Diff S^3)".

(13) François Laudenbach : Topologie de la dimension trois : homotopie et isotopie, Astérisque 12 (1974), Soc. Math. de France, 1974.

(14) François Laudenbach et Oscar Zambrano : Disjonction à paramètres dans une variété de dimension 3, P^2-irréductible, C.R. Acad. Sc. Paris, t. 288 (5 février 1979), Série A, 331-334.

(15) John W. Milnor : On manifolds homeomorphic to the 7-spheres, Annals of Math. 64 (1956), 399-405.

(16) John W. Milnor : Morse Theory, Annals of Mathematics Studies, Princeton University Press, #51.

(17) John W. Milnor and James D. Stasheff : Characteristic classes, Annals of Mathematics Studies, Princeton University Press, # 76, 1974.

(18) Bernard Morin et Jean-Pierre Petit : Problématique du retournement de la sphère, C.R. Acad. Sc. Paris, t. 287 (23 octobre 1978), Série A. 767-770.

(19) Bernard Morin et Jean-Pierre Petit : Le retournement de la sphère, Pour la Science, n°15, janvier 1979, p. 34 à 49.

(20) M. Morse and E. Baiada : Homotopy and homology related to the Schoenflies problem, Annals of Math. Series 2, t. 58, 1953, p. 142-165.

(21) S.P. Novikov : Differentiable sphere bundles, Izv. Aka. Nauk, SSSR Mat 29 (1965), 1-96 (Am. Math. Soc. Translations, Ser. 2, Vol. 63).

(22) Anthony Phillips : Turning a surface inside out, Scientific American, n° 214, pp. 112-120, May 1966.

(23) Stephen Smale : A classification of immersions of the two-sphere, Trans. Amer. Math. Soc., 90, 1959, p. 281-290.

(24) Stephen Smale : Diffeomorphisms of the 2-spheres, Proc. Amer. Math. Soc. 10 (1959), 621-626.

LOCAL CHEVALLEY COHOMOLOGIES OF THE DYNAMICAL LIE ALGEBRA OF A SYMPLECTIC MANIFOLD

M. De Wilde

University of Liège, Belgium

0. INTRODUCTION

Let (M,F) be a connected symplectic manifold. We denote by N the space of all smooth functions of M, equipped with the Poisson bracket, by L (resp. $L^\star$) the space of all locally (resp. globally) hamiltonian vector fields on M, equipped with the Lie bracket. Recall that, if $(G,[\ ,\ ])$ is a Lie algebra and (F,ρ) a representation of G, the corresponding Chevalley cohomology $H(G,\rho)$ is the cohomology of the complex

$$\ldots \to \Lambda^p(G,F) \overset{\partial}{\to} \Lambda^{p+1}(G,F) \to \ldots$$

where $\Lambda^p(G,F)$ is the space of p-linear alternating maps from $G \times \ldots \times G$ into F and

$$\partial C(u_o \ldots u_p) = \sum_{i \leqslant p} (-1)^i \rho(u_i) C(u_o, \ldots \hat{u}_i \ldots, u_p)$$

$$+ \sum_{i<j} (-1)^{i+j} C([u_i, u_j], u_o, \ldots \hat{u}_i \ldots \hat{u}_j \ldots, u_p) ,$$

where $\hat{X}$ means that X is omitted. We restrict here our attention to the above mentionned Lie algebras and their adjoint representation. A further restriction consists in taking not all p-cochains C but only the local ones (By Peetre' theorem (2), these are precisely the multilinear differential operators). The corresponding

M. Cahen et al. (eds.), Differential Geometry and Mathematical Physics, 131–136.

cohomology spaces are denoted $H_{loc}(N)$, $H_{loc}(L)$, $H_{loc}(L^{\star})$.

The relation between the analoguous cohomology spaces, where the cochains are differential operators of order 1 (denoted $H_{1-diff}(N),\ldots$) is given by A. Lichnerowicz in (6). We extend here his results to the local cohomologies. The basic step in this extension is the rather surprising property (prop. 2.1) : if the range of a differential operator L lies in the space of closed p-forms ($p < \dim M$) of a manifold, then it lies in the space of exact p-forms and L factorizes into $d \circ L'$, where L' is another differential operators.

The details of the proofs are given in (3).

1. A DIVISION LEMMA

Let E be a finite dimensional vector space; we denote by $A^p(E)$ the space of covariant alternating p-tensors on E and by $P^p(E)$ the space of polynomials on $E^{\star}$ with coefficients in $A^p(E)$. The set of elements of $P^p(E)$ without constant terms is denoted by $P^p_{\star}(E)$. We consider the map $\partial_p : P^p(E) \to P^{p+1}(E) : \omega \to \xi \wedge \omega$.

Proposition 1.1. *For* $1 \leqslant p < \dim E$, *there exists a linear map* $k_p : \ker \partial_p \to P^{p-1}(E)$ *such that* $\partial_{p-1} \circ k_p = id_{\ker \partial_p}$. *For* $p = \dim E$, *there exists a linear map* $k_p : \operatorname{im} \partial_{p-1} = P^p_{\star}(E) \to P^{p-1}(E)$ *such that* $\partial_{p-1} \circ k_p = id_{P^p_{\star}(E)}$.

The proof is based on elementary algebraic observations.

Corollary 1.2. *The cohomology of the complex* $\ldots \to P^p(E) \xrightarrow{\partial_p} P^{p+1}(E) \to \ldots$ *is given by* $H^p_{\partial} = 0$, *for* $p \neq \dim E$; $H^p_{\partial} \cong A^p(E) \cong \mathbb{R}$ *for* $p = \dim E$.

2. FACTORIZATION PROPERTY FOR LOCAL MAPS

In this paragraph, M denotes a Hausdorff, second countable manifold, E a vector bundle on M; $\Gamma(E)$ is the space of smooth sections of E and $\Lambda^p(M)$ [resp. $B^p(M), Z^p(M)$] is the space of smooth (resp. closed, exact) p-forms on M.

Proposition 2.1. *Let* $L : \Gamma(E)^q \to \Lambda^p(M)$ *be a* q*-linear local map. Assume that, for all* $\gamma_i \in \Gamma(E)$, $L(\gamma_1,\ldots,\gamma_q)$ is a closed p-form. Then,

(i) *if* $p = 0$, $L = 0$;

(ii) *if* $1 \leqslant p < \dim M$, *there exists a* q*-linear local map* $L' : \Gamma(E)^q \to \Lambda^{p-1}(M)$ *such that* $L = d \circ L'$. *In particular, whenever the range of* L *lies in* $Z^q(M)$, *it lies in* $B^q(M)$;

(iii) *if* $p = \dim M$, $L = L_o + dL'$, *where* $L' : \Gamma(E)^q \to \Lambda^{p-1}(M)$ *is a* q*-linear map and* $L_o : \Gamma(E)^q \to \Lambda^p(M)$ *a unique local map of order* 0 *in the last argument, i.e., such that* $L_o(\gamma_1,\ldots,\gamma_q)$ is $C_\infty(M)$*-linear with respect to* γ_q *for all* $\gamma_1,\ldots,\gamma_{q-1} \in \Gamma(E)$.

In a domain of chart, the problem is reduced to the algebraic problem of §1 by passing to the symbol of L and by an induction on the degree of L.

Extending the property to an arbitrary manifold is then a matter of glueing the L''s obtained in the first part of the proof.

Denote by $A^p(M)$ the bundle of covariant p-tensors on M and by $L_q(E,F)$ the space of all q-linear local maps from $\Gamma(E)^q$ into $\Gamma(F)$, where E,F are two vector bundles on M.

Corollary 2.2. *If* $\tilde{d}$ *is the coboundary operator* $\tilde{d}L = d \circ L$, *the cohomology of the complex* $(L_q(E,A^p(M)),d)$ *is given by* $H^p_{q,d} = 0$ *for* $p \neq \dim M$; $H^p_{q,d} \cong L_{q-1}(E,E^\star \otimes A^p(M))$ *for* $p = \dim M$.

It follows immediately from Proposition 2.1., observing that the elements of $L_q(E,A^p(M))$, of order 0 in the last argument identify in a natural way to the elements of $L_{q-1}(E,E^\star \otimes A^p(M))$.

Remarks 2.3. (i) *If, in proposition* 2.1.*, (i) and (ii),* L *is symmetric, or alternating, we may choose* L' *symmetric or alternating, we have just to symmetrize or antisymmetrize* L'.

The results of corollary 2.2., for $p \neq \dim M$, extend thus straightforwardly to symmetric or alternating maps.

Let us call a *differential operator* a local operator which is of finite order all over the manifold.

(ii) If L *is a differential operator, so are* L' *and* L_o : indeed, the degree of L' and L_o never exceeds the degree of L.

3. THE LOCAL CHEVALLEY COHOMOLOGY OF $L^{\star}$ AND L

Let (M,F) be a connected symplectic manifold. We denote by μ the isomorphism $x \to i(X)F$ of the space of smooth vector fields of M onto the space of smooth 1-forms of M.

Given a local p-cochain C of N (i.e. an alternating local map of N^p into N), consider the diagram

$$\begin{array}{ccc} (X_{f_o},\ldots,X_{f_{p-1}}) & \overset{C'}{\dashrightarrow} & C'(X_{f_o},\ldots,X_{f_{p-1}}) \\ \uparrow {\scriptstyle (\mu^{-1}\circ d,\ldots,\mu^{-1}\circ d)} & & \uparrow {\scriptstyle \mu^{-1}\circ d} \\ (f_o,\ldots,f_{p-1}) & \overset{C}{\longrightarrow} & C(f_o,\ldots,f_{p-1}) \end{array}$$

and try to define C' by asking that the diagram commutes.

(i) If C is constant whenever one of the functions f_i is constant, C' is well defined. It is then clearly local. Consider now the map $j : C \to C'$ between the space of local cochains of N, constant on the constants, and the local cochains of $L^{\star}$. It is bijective : injectivity is almost obvious; surjectivity follows from Proposition 2.1. applied to $L : (f_o,\ldots,f_{m-1}) \to C'(X_{f_o},\ldots,X_{f_{p-1}})$. Moreover, if ∂ and δ are the coboundary operators, $\delta C'$ corresponds to ∂C.

If we denote by $H_{loc}, H_{diff}, H_{p\text{-}diff}$ the cohomology spaces corresponding to the local, differentiable or 1-differentiable cochains and by $H_{loc,cc},\ldots$ the corresponding cohomologies for the cochains which are constant on the constants, we have

<u>Proposition 3.1</u>. *The space* $H_{loc}(L^{\star})$, $H_{diff}(L^{\star})$, $H_1(L^{\star})$ *are respectively isomorphic to* $H_{loc,cc}(N)$, $H_{diff,cc}(N)$ *and* $H_{1,cc}(N)$.

The case of the 1-differentiable cohomologies is given by A. Lichnérowicz in (6), where $H_{1,cc}(N)$ is explicitly computed.

The spaces $H^2_{diff}(N)$, $H^3_{diff}(N)$ are computed by S. Gutt in (4) and their equivalence with the corresponding local cohomology spaces is given in (2).

If $H_{\ldots,nc}$ denotes the cohomology space corresponding to the cochains which vanish when one of the arguments is constant, the spaces $H^2_{diff,nc}(N) \cong H^2_{loc,nc}$ and $H^3_{diff,nc} \cong H^3_{loc,nc}$ are also given in (2,4).

The relation between $H_{loc,cc}$ and $H_{loc,nc}$ is easy to describe.

Let C be a local p-cochain which is constant when one of the arguments is constant. Taking its local form in any local coordinate system, we see that a non-zero constant term necessarily comes from a term of degree zero in only one argument. Such a term does not exist unless C is a 1-cochain.

It is thus straightforward that

$$H^p_{loc,nc}(N) = H^p_{loc,cc}(N) \text{ and } H^p_{diff,nc}(N) = H^p_{diff,cc}(N)$$

for $p > 2$.

If C is a local 1-cocycle, i.e. a derivation of N, it is shown in (1) that

$$C(u) = \mathcal{L}_X u + k(X)u$$

where $X \in L^c$ and L^c is the set of all vector fields such that $\mathcal{L}_X F + k(X)F = 0$ ($k(X) \in \mathbb{R}$). It is thus obvious that

$$H^1_{loc}(N) \cong H^1_{diff}(N) \cong H^1_{loc,cc}(N) \cong H^1_{diff,cc}(N) \cong L^c/L^\star .$$

Let now C be a local (or differentiable) 2-cocycle. It is shown in (4,5) that $C = S^3_\Gamma + \delta E + A$, where $\alpha \in \mathbb{R}$, S^3_Γ is a non exact 2-cocycle, vanishing on the constants, E a 2-cochain, A a 1-differentiable cocycle. It follows then as in (5) that

$$H^2_{loc,cc}(N) \cong H^2_{diff,cc}(N)\ \mathbb{R}\ S^3 \oplus H^2_{1\text{-}diff,cc}(N) .$$

By (6),

$$H^2_{1\text{-}diff,cc}(N) \cong H^2(M, \mathbb{R}) \,/\, \mathrm{im}\, F^o_\#$$

where $F^p_\# : H^p(M, \mathbb{R}) \to H^{p+2}(M, \mathbb{R})$ is the map induced by $u \to F \wedge u$. Hence

$$H^2_{loc,cc}(N) \cong H^2_{diff,cc}(N) \cong \mathbb{R}\, S^3_\Gamma \oplus [H^2(M, \mathbb{R})/\text{im } F^o_\#].$$

The results for the local Chevalley cohomology of L are the following

Proposition 3.2. *The following spaces are isomorphic : for* $p < 1$, $H^p_{loc}(L) \cong H^p_{loc}(L^\star) \cong H^p_{diff}(L) \cong H^p_{diff}(L^\star)$; $H^p_{1\text{-}diff}(L) \cong H^p_{1\text{-}diff}(L^\star)$.

For $p = 1$,

$$H^1_{loc}(L) \cong H^1_{diff}(L) \cong H^1_{1\text{-}diff}(L) \cong L^c/L.$$

The result about the 1-differentiable cohomologies are already in (6).

REFERENCES

(1) A. Avez, A. Lichnerowicz, A. Diaz-Miranda : Sur l'algèbre des automorphismes infinitésimaux d'une variété symplectique. J. Diff. Geometry 9(1974), pp. 1-40.

(2) M. Cahen, M. De Wilde, S. Gutt : Local cohomology of the algebra of C^∞ functions on a connected manifold. Lett. in Math. Phys. 4(1980), pp. 157-167.

(3) M. De Wilde : On the local Chevalley cohomology of the dynamical Lie algebra of a symplectic manifold. Lett. in Math. Phys. 5(1981), pp.351-358.

(4) S. Gutt : Second et troisième espaces de cohomologies différentiables de l'algèbre de Lie de Poisson d'une variété symplectique. Ann. Inst. H. Poincaré 33(1980), pp. 1-31.

(5) S. Gutt : Déformations formelles de l'algèbre des fonctions différentiables sur une variété symplectique. Doctor thesis, Univ. of Brussels (1980).

(6) A. Lichnerowicz : Cohomologie 1-différentiable des algèbres de Lie attachées à une variété symplectique ou de contact. J. Math. Pures et Appl. 53(1974), pp.459-484.

THE SPIN POLARIZER

C. Duval

Centre de Physique Théorique CNRS Marseille, France

FOREWORDS

This communication is to be considered as an addendum to the reference (1). The purpose of that paper was to investigate, within the framework of geometric quantization à la Kostant - Souriau, the new notion of polarizer of a prequantizable symplectic manifold. Although that work was mainly concerned with the general scheme of compact semi-simple Lie groups, its ultimate motivation was clearly of physical nature. That is why it was there proposed to examine polarizers; a drastic selection of prequantizable coadjoint orbits showed up, which, in fact, might correspond to the actual selection of physically relevant multiplets of hadron spectroscopy. The question then arises : must one take polarizers seriously? In order to strenghthen our point of view, we propose here to discuss the properties of polarizers for spinning particles. The non relativistic Dirac particle, its Poincaré invariant polarizer is explicitely worked out, as well as for the spin 1 relativistic massive particle. The associated mixed (real + Kähler) polarization naturally corresponds to the unique Poincaré invariant polarization discovered by Souriau and Renouard.

M. Cahen et al. (eds.), Differential Geometry and Mathematical Physics, 137–147.

I. THE NOTION OF POLARIZER OF A PREQUANTIZABLE DYNAMICAL SYSTEM

Let (U,σ) be a symplectic manifold (σ is a closed 2 form with trivial kernel), we shall say that (U,σ) is prequantizable if there exists a principal circle bundle (Y,ω) over it with a connection $i\,\omega$ whose curvature $/i$ is the pull - back of the symplectic structure σ of U. If the manifold U is simply connected, all prequantizations (Y,ω) are "equivalent" and one can thus speak of the prequantization of the symplectic manifold.

For an extensive account on that subject, one can consult the pioneering references of the so called KOSTANT - SOURIAU theory of geometric quantization (3) (6), also (8) (9).

One of the most remarkable features of prequantization is to provide a necessary link half-way between a classical dynamical system and its quantization. It originally helped for a better understanding of spin quantization for example, as well as other related physical phenomena such as Dirac monopole quantization. Let us illustrate that situation by the non trivial prequantization of the 2 - sphere (S^2, s surf) [1] which can be considered as the "spin phase space" of a non relativistic particle of spin s (3) or as the "internal isospin phase space" of a WONG particle (4) (5) (2), etc...

Since $S^2 \cong \mathbb{P}_1(\mathbb{C})$ is simply connected, it will be sufficient to exhibit one prequantization; the Hopf fibration $S^3 \to S^2$ will do it as follows.

Let us start with the principal bundle $(S^3, 2s\,\bar{z}\,dz/i)$ over S^2 : here $z \in \mathbb{C}^2$, $\bar{z}\,z = 1$ (the bar "-" denotes the canonical hermitian structure of $\mathbb{C}^2$) and the projection $p : S^3 \to S^2$ is given by $p(z) = 2s\,(z\,\bar{z} - 1/2)/i = s\,\sigma(v)/i$ where $\sigma(v) := \sigma_A\, v^A$ (the σ_A's - A = 1,2,3 - denote the Pauli matrices) and $v \in S^2 \hookrightarrow \mathbb{R}^3$. Furthermore z, z' lie in the same fibre i. e. $z \sim z'$ iff $z = z'u$ ($u \in U(1)$).

(S^2 , s surf) turns out to be in fact a (co)adjoint orbit (with Casimir $s > 0$) of $SU(2) \cong S^3$, and the preceding construction is nothing else that the coadjoint orbit setting (1). As for the (real) 1 - form

$$\omega := 2s\ \bar{z}\ dz/i \tag{1.1}$$

of S^3, it is easy to show that

$$d\,\omega = p^{\star}\ (\sigma) \tag{1.2}$$

where σ = s surf.

Now i ω is almost a connection form on S^3, but $\omega(\xi) = 2s$ ($\neq 1$ in general) ($\xi := i\ z\ \frac{\partial}{\partial z}$ denotes the U(1) generator on S^3); suppose that 2s is an integer, so that one may consider the quotient $Y_3 := S^3/\mathbb{Z}_{2s}$.

It is not hard to prove that ω is the pull - back by the projection $S^3 \to Y_3$ of a connection form of Y_3 whose curvature is the pull - back of σ by the projection $Y_3 \to S^2$. We have thus achieved the complete prequantization of the phase spaces for those spins such that

$$2s \in \mathbb{N} \tag{1.3}$$

Note that condition (1.3) is a particular case of the general criterion for the prequantazibility of a symplectic manifold (U,σ), namely $1/2\pi \int_{S^2} \sigma \in \mathbb{Z}$ for every 2 - cycle S^2 of U (σ defines a class [σ] in H^2 (U, $\mathbb{Z}$) (6).

The preceding construction can be generalized to the case of coadjoint orbits of semi-simple and simply connected compact Lie groups. The physical interpretation of these results is discussed in (2).

Another important notion in the theory of geometric quantization is the notion of polarization of a symplectic manifold. Physically speaking choosing a polarization amounts to making the choice of a representation (e.g. the "p" or "q" picture of quantum mechanics). One happens to have a fairly narrow choice in some special cases, for instance in the case of homogeneous symplectic manifolds where one is led to find out all invariant polarizations. This is in fact the content of KIRILLOV's method of coadjoint orbits (7). As an important result, let us quote the Borel-Weyl-Bott theorem which establishes the one to one correspondence between unitary irreducible representations of compact Lie groups and their prequantizable (co)adjoint orbits. The theorem still works in the case of nilpotent and certain solvable Lie groups (7) (10).

Let us now introduce the definitions of a polarization and also of more particular geometrical objects called polarizers (SOURIAU). It has been argued in (1) that invariant polarizers (which provide us with invariant polarizations) might be of some physical interest since they give rise to a strong selection of homogeneous dynamical systems corresponding to actual physical situations. A polarization F of a symplectic manifold (U,σ) is, by definition, an involutive distribution of maximal isotropic subspaces $F_x \subset T_x^{\mathbb{C}} M$; i.e. $\mathbb{C}\text{-dim}(F_x) = 1/2 \dim(U)$, $\sigma(F_x,F_x)=0^{\,2}$. If (Y,ω) is a prequantization of (U,σ), F can be horizontally lifted to Y, and one thus gets a Planck-polarization $\tilde{F}$ (see (3) for the real case). Examples of real, Kähler, mixed polarizations can be found in (3) (8) (9).

We will now produce an alternative definition of polarizers which have been originally introduced in (1).

We shall call polarizer every complex n - form ϕ of (Y_{2n+1},ω) such that

$$\begin{aligned} &\text{i)}\ \phi \text{ is horizontal} \\ &\text{ii)}\ \text{rank}(\phi) = n \\ &\text{iii)}\ u^{*}\phi = u^{k}\phi \qquad \forall\, u \in U(1) \quad (k \in \mathbb{Z} - \{0\}) \\ &\text{iv)}\ D\phi = 0 \end{aligned} \tag{1.4}$$

According to the current terminology (11), ϕ is a complex - valued horizontal n-form of type ρ_k $(\rho_k(u) = u^k)$ with minimal rank and vanishing covariant derivative. The last condition (iv) can be written, using the definition $D\phi := d\phi \circ \text{hor}$ (where hor denotes the horizontal projection).

$$D\phi = d\phi - i\,k\,\omega \wedge \phi = 0 \tag{1.5}$$

Equation (1.5) is called the polarizer equation.

It has already been shown that $\ker(d\phi)$ is a Planck-polarization associated with the polarizer ϕ (1).

Since we focus, in this paper, our attention on the Kähler

case, we will give the standard example of SU(2) invariant polarizer of the spin phase space (see above). It can be proved with the help of techniques described in (1), that

$$\phi = \text{vol } (z) \tag{1.6}$$

is the only invariant SU(2) polarizer on $(S^3, 2s\,\bar{z}\,dz{:}i)$ [3]. We denote by vol the canonical volume element of $\mathbb{C}^2$, so that vol(z) is a complex 1-form of S^3 which turns out to be horizontal and of rank 1. As for the condition (1.5), we have $d\,\phi = 2\,\text{vol}$, $i\,\omega \wedge \phi = 2s\,\bar{z}\,dz \wedge \text{vol}\,(z) = 2s\,\text{vol}$ and ϕ is actually a polarizer for those spins such that

$$k\,s = 1 \tag{1.7}$$

The only allowed values for the spin are respectively 1/2 and 1! We have shown that these orbits are associated with the representations {2} and {3} of SU(2). Again, generalizations are possible and some of them are treated in (1) together with a discussion on the physical implications of the requirement of the existence of invariant polarizers.

Let $a \in$ Quant (Y,ω) i. e. $a \in$ Aut (Y) and $a^{*}\,\omega = \omega$, if ϕ is a k-polarizer of (Y,ω), so is $a^{*}\,\phi$ (trivial). Suppose now that G is a (Lie) subgroup of Quant (Y,ω) which preserves ϕ conformally i. e.

$$a^{*}\,\phi = f\,\phi \tag{1.8}$$

where f is a nowhere vanishing complex function of Y depending upon $a \in G$, then

$$df \wedge \phi = 0 \tag{1.9}$$

The proof goes as follows : $d(a^{*}\phi) = a^{*}(d\,\phi) = a^{*}(i\,k\,\omega \wedge \phi) = i\,k\,\omega \wedge a^{*}\,\phi$ (see (1.5)). Thus $d(f\,\phi) = df \wedge \phi + f\,d\phi = df \wedge \phi + i\,k\,f\,\omega \wedge \phi = i\,k\,f\,\omega \wedge \phi$ (see above). Whence the result (1.9).

Since $\tilde{F} := \ker(d\,\phi)$ [4] is a Planck polarization associated with the polarizer ϕ, $a_{*}\,\tilde{F}$ is another Planck polarization associated with $(a^{-1})^{*}\,\phi$. Furthermore (1.9) implies that

$$d(a^{*}\,\phi) = f\,d\,\phi \qquad (a \in G) \tag{1.10}$$

and thus $\ker(d\,(a^{*}\,\phi)) = \ker\,(d\,\phi)$. At last

$$a_{\star} \tilde{F} = \tilde{F} \qquad \forall a \in G \tag{1.11}$$

If ϕ is a conformally invariant polarizer with respect to a Lie subgroup of Quant (Y,ω), the associated Planck polarization (hence the polarization) is G invariant. The converse is true and easy to prove.

Examples of strictly invariant polarizers are introduced in (1). The next section is devoted to the case of a conformally Poincaré invariant polarizer for a relativistic massive spinning particle.

II. THE POINCARE INVARIANT POLARIZER OF THE RELATIVISTIC MASSIVE SPINNING PARTICLE

The canonical presymplectic structure on the universal covering $\hat{P}_{10}$ of the neutral component P_{10} of the Poincaré group $O(3,1) \underline{x} \mathbb{R}^{3,1}$ that leads to the geometric description of the prequantum model of a massive spinning particle has already been introduced in the reference (2). It has also been emphasized that $\hat{P}_{10}$ itself turns out to be a prefered arena for the geometry which is necessary to work out the canonical symplectic structure of the space motions (a coadjoint orbit $O_{s,m}$ of P_{10} with Casimirs $s > 0$ (the spin) and $m > 0$ (the mass)) and also the (unique) prequantization of $O_{s,m}$. Let us recall the following result (3):

the space of motions of a particle of spin s and mass m is prequantizable if the spin is quantized according to the prescription

$$2s \in \mathbb{N} \tag{2.1}$$

just as in the non relativistic case (1.3).

In order to find the invariant polarizers of the prequantum bundle of that model, we need only the following construction which is due to SOURIAU and can be found in (3); see also (5).

Let us start with the "prequantum <u>evolution space</u>".
$W_{10} := \{(X,\zeta) \in M \times \mathbb{C}^{2,2}\ ,\ \ \bar{\zeta}\,\zeta = 1,\ \bar{\zeta}\ \ \gamma_5\zeta = 0\ \}$ equipped with the (real) 1-form

$$\bar{\omega} := 2s\ \bar{\zeta}\ d\ \zeta/i\ \ + m\ \bar{\zeta}\ \gamma(dX)\zeta \tag{2.2}$$

where (M,g) denotes the Minkowski space-time ($g = \mathrm{diag}(-1,-1,-1,+1)$) and the bar "-" the hermitian structure of $\mathbb{C}^{2,2}$

$$G = 1/2 \begin{pmatrix} 0 & 1 \\ 1 & 0 \end{pmatrix} \tag{2.3}$$

The equations of motion are derived by computing the expression of the distribution ker $(d\bar{\omega})$, so that

$$\delta \begin{pmatrix} \zeta \\ X \end{pmatrix} \in \ker (d\bar{\omega}) \Longleftrightarrow \begin{cases} \delta\zeta = i\alpha\ \zeta \\ \delta X = \beta\ I \end{cases} \tag{2.4}$$

where α, β are two real Lagrange multipliers and I is the unit future pointing vector of M defined by [5]

$$\bar{I}\ V := \bar{\zeta}\ \gamma(V)\zeta \quad \forall V \in M \tag{2.5}$$

(we have used the notation $\bar{I} := g\ (I,.)$).

The quantity $P := m\ I$ is interpreted as the linear momentum and is a constant of the motion, i. e. $\delta\ P = 0$ ((2.4),(2.5)). [6].

As for the <u>space of motions</u> : $U_8 = W_{10}/\ker\ (d\bar{\omega})$, it turns out to have the topology $\mathbb{R}^6 \times S^2$, and to be symplectomorphic to the coadjoint orbit $\mathcal{O}_{s,m}$ of $\hat{P}_{10}$ as $\hat{P}_{10}$ acts on W_{10} according to

$$\begin{aligned} \zeta &\to A\ \zeta \\ X &\to LX + C \end{aligned} \tag{2.6}$$

where $A \in \hat{\mathcal{L}}_6 \cong Sl(2,\mathbb{C})$ ($\mathcal{L}_6$ denotes the neutral component of the Lorentz group).

$$A = S \begin{pmatrix} a & 0 \\ 0 & \bar{a}^{-1} \end{pmatrix} S^{-1} \quad (a \in Sl(2,\mathbb{C})) \tag{2.7}$$

and $C \in M$ (translation)

Let us now introduce the new variables

$$Y := (1 - I\ \bar{I})\ X \tag{2.8}$$

$$Z := \zeta\ e^{i\bar{P}X/2s} \tag{2.9}$$

which describe a manifold $\tilde{Y}_9$ whose quotient by $\mathbb{Z}_{2s}$ [7] ($2s \in \mathbb{N}$) turns out to be <u>the</u> prequantum bundle Y_9 of U_8 since the latter is simply connected (3). In fact $\bar{\omega}$ descends on $\tilde{Y}_9$ as the following 1-form

$$\omega = 2s\ \bar{Z}\ dZ/i - \bar{Y}\ dP \tag{2.10}$$

Since $\hat{P}_{10}$ acts on W_{10} as a group of bundle automorphisms preserving the structure $\bar{\omega}$, so does it on $(\tilde{Y}_9,\omega)$ according to the action

$$Z \rightarrow AZ\, e^{i\bar{C}LP/2s} \tag{2.11}$$

$$Y \rightarrow LY + (1 - LI\bar{L}\bar{I})\, C \tag{2.12}$$

where $L \in \mathcal{L}_6$ corresponds to the element $A \in \hat{\mathcal{L}}_6$ $(\gamma\,(LV) := A\gamma(V)A^{-1})$. We also have

$$P \rightarrow LP \tag{2.13}$$

We now need some information on the algebraic properties of the spinors Z satisfying the constraints $\bar{Z}\, Z = 1$, $\bar{Z}\, \gamma_5\, Z = 0$. It can be proved, using equation (2.5), that

$$\gamma(I)\; Z = Z \tag{2.14}$$

and thus

$$Z = S \begin{pmatrix} z \\ \sigma(P)\; z/m \end{pmatrix} \tag{2.15}$$

where $\sigma(P) := -\sigma_A\, P^A + P^4$ and $z \in \mathbb{C}^2$ satisfies

$$\bar{z}\; \sigma\,(P)\; z = m \tag{2.16}$$

The 1-form ω (2.10) reads now

$$\omega = 2s/mi\; \bar{z}\; \sigma(P)\; dz + s/mi\; \bar{z}\; \sigma(dP)z - \bar{Y}\; dP \tag{2.17}$$

It is now possible to look for all 4 - forms on $\tilde{Y}_9$ that fulfill the condition (1.5) of polarizers and are furthermore $\hat{P}_{10}$ (conformally) invariant. The result goes as follows:
Let Vol(P) denote the Lorentz invariant volume element of the forward mass hyperboloid ($\bar{P}\, P = m^2$) and vol the canonical volume element of $\mathbb{C}^2$; let us look under which conditions the 4-form

$$\phi := \mathrm{vol}(z) \quad \mathrm{Vol}(P) \tag{2.18}$$

can be a polarizer (see (1.6)). We have $d\phi = 2\ \mathrm{vol} \wedge \mathrm{Vol}(P)$, since Vol(P) is closed, and

$i\ \omega \wedge \phi = 2s/m\ \bar{z}\ \sigma(P)\ dz \wedge \mathrm{vol}(z) \wedge \mathrm{Vol}(P) =$

$2s/m\ \mathrm{Tr}(z\ \bar{z}.\sigma(P))\ \mathrm{vol} \wedge \mathrm{Vol}(P) = 2s\ \mathrm{vol} \wedge \mathrm{Vol}(P)$ (see (2.16)).

Hence ϕ is a polarizer if the integer k ((1.4)iii) satisfies

$$ks = 1 \tag{2.19}$$

just as in the non relativistic case (1.7). The conditions (2.1) and (2.19) again imply that the only allowed values of the spin are 1/2 and 1! The polarizer (2.18) turns out to be the unique $\hat{P}_{10}$ conformally invariant (see (2.7), (2.11), (2.13)) one :

$$(A,C)^{☆} \phi = e^{i\bar{C}LP/s} \phi \tag{2.20}$$

It corresponds in fact to the unique invariant polarization on U_8 (RENOUARD (12)). Let us stress that condition (2.19) implies that ϕ passes to the quotient Y_9.

As a final remark, we would like to put the expression (2.18) in a different guise with the help of the quaternionic structure J of $\mathbb{C}^{2,2}$

$$J = S \begin{pmatrix} 0 & j \\ j & 0 \end{pmatrix} S^{-1} \tag{2.21}$$

where j stands for the quaternionic structure of $\mathbb{C}^2$.

A simple calculation shows that $\mathrm{vol}(z) = \overline{jz}\, dz$, and if we define

$$Z^{☆} = \pi Z \tag{2.22}$$

where $\pi := (1 + i\,\gamma_5)/2$ is a null projector of $\mathbb{C}^{2,2}$, then

$$\overline{jz}\, dz = 2\, \overline{JZ^{☆}}\, dZ^{☆} \tag{2.23}$$

At last [8]

$$\overline{JZ^{☆}}\, dZ^{☆} \wedge \mathrm{Vol}(P) \tag{2.24}$$

is the sought Poincaré conformal invariant polarizer of the relativistic massive spinning particle.

1 Surf denotes the canonical surface element of the unit 2-spheres S^2.

2 It is also assumed that $\dim (F_x \cap \bar{F}_x) = \text{const.}$ (admissibility).

3 The polarizer equation (1.5) can be pulled back to S^3, and the form (1.6) of ϕ turns out to descend to Y_3 if certain conditions hold (see below). The crucial point is that polarizers are <u>global</u> geometrical objets.

4 $\ker (d\,\phi) \subset \ker (\phi)$.

5 We use the following representation of Dirac γ - matrices in a spinor frame S

$$\gamma_A = \begin{pmatrix} 0 & \sigma_A \\ -\sigma_A & 0 \end{pmatrix} \quad (A = 1,2,3) \;;\gamma_4 = \begin{pmatrix} 0 & 1 \\ 1 & 0 \end{pmatrix};\gamma_5 = \begin{pmatrix} -i & 0 \\ 0 & i \end{pmatrix}$$

$\bar{S}\, S = G$ (2.3). A shorthand notation $\gamma(V) := \gamma_A V^A + \gamma_4 V^4$

6 The leaves of the characteristic foliation of $d\,\bar{\omega}$ project onto timelike geodesics of spacetime M, the world lines of the particle. Whence the justification of the terminology: "evolution space" for W_{10}.

7 $\{(Y,Z) \sim (Y',Z')\} \iff \{Y = Y' \;;\; Z = Z' u,\ u \in \mathbb{Z}_{2s}\}$
Note that $\tilde{Y}_9 = W_{10}/\ker(\bar{\omega}) \cap \ker(d\bar{\omega})$ which is clear by (2.2) and (2.4).

8 The 1-form (2.23) has also been considered from the different point of view of twistor geometry (13).

REFERENCES

(1) C. DUVAL, On the Polarizers of Compact Semi-Simple Lie Groups. Applications. Ann. Inst. Henri Poincaré A, vol XXXIV, n°1 (1981).

(2) C. DUVAL, On the Prequantum Description of Spinning Particles in an External Gauge Field, in : Méthodes de Géométrie Différentielle en Physique - Mathématique. Coll. Int. C.N.R.S. Proceedings Aix en Provence 1979, Springer Verlag 836, Berlin (1980).

(3) J.M. SOURIAU, Quantification Géométrique. Comm. Math. Phys. Vol 1 (1966) - Structure des Systèmes Dynamiques. Dunod, Paris (1969) - Structure of Dynamical Systems (In preparation).

(4) S. STERNBERG, On the Role of Field Theories in our Physical Conception of Geometry. Differential Méthods in Mathematical Physics II. Proceedings Bonn (1977), Springer Verlag 676, Berlin (1978).

(5) S. STERNBERG, T. UNGAR, Classical and Prequantized Mechanics without Lagrangians or Hamiltonians. Hadr. Journal, vol 1, N° 1, Nonatum Press Mass. (1978).

(6) B. KOSTANT, Quantization and Unitary Representations. Lect. Notes in Math, Springer Verlag 170, Berlin (1970).

(7) A. KIRILLOV, Elements de la Théorie des Représentations. Ed Mir, Moscou (1974).

(8) D.J. SIMMS, N.M.J. WOODHOUSE, Lecture on Geometric Quantization. Lect. Notes in Physics, Springer Verlag 53, Berlin (1976).

(9) J. SNIATYCKI, Geometric Quantization and Quantum Mechanics. Springer Verlag, New York (1980).

(10) A. WEINSTEIN, Symplectic Geometry. Preprint Univ. of California, Berkeley (1981).

(11) S. KOBAYASHI, K. NOMIZU, Foundations of Differential Geometry I. Interscience, New York (1962).

(12) P. RENOUARD, Variétés Symplectiques et Quantification; thèse, Orsay (1969).

(13) R.O. WELLS, Jr., Complex Manifolds and Mathematical Physics. Bull. Am. Math. Soc. Vol 1, n°2 (March 1979).

VARIATIONAL THEORY IN FIBRE BUNDLES : EXAMPLES

J. Eells

Mathematics Institute, University of Warwick.

I. GENERAL REMARKS

Nature is uncompromisingly efficient. That article of faith has shaped much of our physical and geometric thinking; and any mathematical or physical model of Nature must display prominently its essential features. Indeed, systematic development of such guidelines into sound mathematical theory has -in large measure- led to one of our most powerful tools: *The variational principle.*

The main purpose of my lecture today is two-fold:

(1) To recall a reasonable general context for variational theory, as given in (5);

(2) to illustrate that context with several examples.

These examples are variously interrelated, both in their physical origins and in their mathematical properties. (For instance, there is a popular analogy between the variational principles of Yang-Mills fields over 4-dimensional manifolds (Example 3 below) and of harmonic maps of surfaces (the so-called chiral models; Example 4 bis below).)

Consequently it is especially enlightening to view them in a unified framework.

For a start : *What are the optional maps* $\phi : M \to N$ *between*

M. Cahen et al. (eds.), Differential Geometry and Mathematical Physics, 149–158.

two spaces ?

In order to have a significant problem, we should specify (a) a prescribed class S of maps. E.g., a homotopy class (i.e., a component of the space $C^o(M,N)$ of all continuous maps $M \to N$); a cobordism class; an embedding class;
(b) a function $F : S \to \mathbb{R}$ whose critical points are the best maps. In particular, we should provide a suitable definition of *critical point*. And of course we would like to insure that F has a critical point in S .

Certainly both F and S should be firmly conditioned by further structures on M,N. In the present context we shall require that they be smooth manifolds, and endowed with various tensorial structures; e.g., smooth volume forms, Riemannian metrics; complex, symplectic, contact structures.

II. THE FRAMEWORK.

Let

$$\gamma : W \to M$$

be a smooth fibre bundle. The k-jets of sections of γ defined in a neighborhood of a point $x \in M$ (= equivalence classes of germs of sections at x, having the same differentials $d^i\phi(x)$ for $0 \leqslant i \leqslant k$) form a smooth manifold $J^k(\gamma)$, and we have the natural fibre bundle maps

$$J^k(\gamma) \to J^{k-1}(\gamma) \to \dots \to J^0(\gamma) = W \to M.$$

Let $C^k(\gamma)$ denote the space of C^k-sections of γ (possibly void). Then $\phi \in C^k(\gamma)$ determines a section

$$j^k(\phi) : M \to J^k(\gamma),$$

which assigns to each point $x \in M$ the k-jet $j^k_x(\phi)$ of ϕ .
A $k^{\underline{th}}$ *order variational density* (or *Lagrangian*) *on* γ is a function

$$f : J^k(\gamma) \to \mathbb{R}.$$

Choose a space $S(\gamma)$ of k-differentiable sections of γ . And fix a smooth positive finite measure ν on M. Define the functional

$$F : \mathcal{S}(\gamma) \to \mathbb{R}$$

by $$F(\phi) = \int_M f(j_x^k(\phi))\nu.$$

Now, we have the notion of *minimum point of* F *in* $\mathcal{S}(\gamma)$, and that should be admitted as a critical point. Secondly, if we introduce a topology on $\mathcal{S}(\gamma)$ then we have

(a) the notion of local minimum of F in $\mathcal{S}(\gamma)$;

(b) components of $\mathcal{S}(\gamma)$. That is important, for we can also ask whether each component contains a critical point. Furthermore, we have Weierstrass' Theorem: If $F : \mathcal{S}(\gamma) \to \mathbb{R}$ is lower semi-continuous, then F achieves a minimum on each compact subset of $\mathcal{S}(\gamma)$.

Remark. The interplay between continuity properties of F and compactness in $\mathcal{S}(\gamma)$ is one of the primary problems of classical calculus of variations. For instance, if $\mathcal{S}(\gamma)$ is a Hilbert space with its weak topology and F is the norm on $\mathcal{S}(\gamma)$ then

(1) F is weakly lower semi-continuous;

(2) closed bounded subsets of $\mathcal{S}(\gamma)$ are compact.

I won't pursue that line of development here, for it is inadequate for many of the variational problems we wish to consider. For instance, in Example 4 bis below, *if either* M *or* N *is an* n-*sphere* ($n \geq 3$) *with any metric, then any harmonic map* $\phi : M \to N$ *which is a minimum of* E *in its homotopy class must be a constant map*. Thus many components will be without minima of E.

An acceptable notion of critical point ϕ_o of $F : \mathcal{S}(\gamma) \to \mathbb{R}$ is the following :

For any germs of 1-parameter deformation $(\phi_t) \subset \mathcal{S}(\gamma)$ of ϕ_o , we require

$$\left.\frac{dF(\phi_t)}{dt}\right|_{t=0} = 0.$$

Then under mild regularity hypotheses, that condition becomes the Euler-Lagrange equation associated with F. In charts that latter takes the form

$$\sum_{|\alpha|\leq k} (-1)^{|\alpha|} \frac{\partial^{|\alpha|}}{\partial x^{\alpha}} (f_{p^{\alpha}}) \equiv 0 \text{ at } \phi_o .$$

Remark. A slightly different version of our basic frame-work is the following:

A kth *order variational density on* γ is a fibre-preserving map

$$\begin{array}{ccc} J^k(\gamma) & \xrightarrow{f} & \Lambda^m T^*(M) \\ & \searrow \quad M \quad \swarrow & \end{array}$$

The *Euler-Lagrange operator* $\mathcal{E}$ of f assigns to each local section ϕ an m-form $\mathcal{E}(j^{2k}\phi)$ on M with values in $T^*(W)$; more precisely, $\mathcal{E}(j^{2k}\phi)$ assigns to $x \in M$ an m-form on W which is basic and has $T^*(W)$ values.

For any compact oriented domain $M_0 \subset M$

$$\frac{d}{dt} \int_{M_0} f(j\phi_k) \Big|_0 = \int_{M_0} \mathcal{E}(j^{2k}\phi)\left(\frac{d\phi_t}{dt}\right) \Big|_0$$

A critical point is a section ϕ such that $\mathcal{E}(j^{2k}\phi) \equiv 0$:

$$\begin{array}{ccc} J^{2k}(\gamma) & \xrightarrow{\mathcal{E}} & T^*(W) \otimes \Lambda^m T^*(M) \\ & \searrow \quad M \quad \swarrow & \end{array}$$

Thus

$$\begin{array}{ccc} C(J^{2k}\gamma) & \rightarrow & C(T^*(W) \otimes \Lambda^m T^*(M)) \\ j^{2k} \nwarrow & & \nearrow \mathcal{E} \\ & C(W) & \end{array}$$

displays $\mathcal{E}$ as a $2k^{th}$ order diferential operator on γ.

Remark. Again, the interplay between the (differentiable) structure of F and the structure of $S(\gamma)$ is a primary problem - the art of the matter! We should have completeness of $S(\gamma)$ with respect to a canonically defined metric. In some favourable situations $S(\gamma)$ can be endowed with a smooth manifold structure with respect to which $F : S(\gamma) \rightarrow \mathbb{R}$ is a smooth function. In such cases the critical points of F are those ϕ_o for which the differential $dF(\phi_o) = 0$. (For instance, see Examples 1 and 4 bis below.)

III. EXAMPLES.

Examples 1. Suppose that $\gamma : W \to M$ is a Riemannian vector bundle over a compact Riemannian manifold. Let A be a linear $k^{\underline{th}}$ order elliptic differential operator on sections of W (with values in some other Riemannian vector bundle over M). Define $S(\gamma) = \mathcal{L}_k^2(\gamma)$, the Sobolev space of sections whose differentials of order $\leqslant k$ are square integrable. Let $j_x^k(\phi) = \frac{1}{2}|(A\phi)(x)|^2$, so

$$F(\phi) = \frac{1}{2}\int_M |(A\phi)(x)|^2 \nu_g,$$

where ν_g is the volume element of the Riemannian metric g on M. Then

$$F : \mathcal{L}_k^2(\gamma) \to \mathbb{R}$$

is weakly lower semi-continuous on the Hilbert space $\mathcal{L}_k^2(\gamma)$. Its Euler-Lagrange equation is $A^{\star}A(\phi) = 0$, where $A^{\star}$ denotes the L^2-adjoint of A. *The extremals of* F *are the sections in the kernel of* A .

Example 2. Let $\pi : P \to M$ be a principal G-bundle, where G is a Lie group. If L(G) denotes the Lie algebra of G and we let G act on it by the adjoint action, then we can form $L(P) = P\times_G L(G)$. Now G also acts on the tangent vector bundle T(P), and we have the following exact sequence of vector bundles

$$0 \to L(P) \to T(P)/G \overset{\pi}{\to} T(M) \to 0;$$

see (1,9). The connexions on the principal bundle π can be characterised as the 1-forms $w \in C(T^{\star}(M) \times T(P)/G)$ such that $\pi \circ w = I$, the identity operator $(I \in C(T^{\star}(M) \times T(M))$. We denote the covariant differential of w by D^w; therefore the curvature of w is $D^w(w)/2$. Relative to suitable Riemannian structures we can define the variational density $f(j^1 w) = \frac{1}{2}\|D^w(w)\|^2$ on the vector bundle

$$\gamma : W = T^{\star}(M) \times T(P)/G \to M.$$

Take $S(\gamma) = \mathcal{L}_1^2(\gamma)$. *The extremals of*

$$F(w) = \frac{1}{2}\int_M \|D^w(w)\|^2 \nu_g$$

are the Yang-Mills fields of $\pi : P \to M$.

Example 3. Suppose that M is compact and oriented. Let $GL^+(M) \to M$ denote its principal bundle of oriented frames. Let m = dim L, and

$$\gamma : W \to GL^+(M)/SO(m) \to M$$

the associated bundle whose fibre is the homogeneous space $GL^+(\mathbb{R}^m)/SO(m)$. The sections of γ are the Riemannian metrics g on M. Let $S(\gamma)$ denote the space of all g such that

$$\int_M \nu_g = 1,$$

of correct Sobolev class (dictated by the functional F_σ below). For any $g \in S(\gamma)$ let

$$R_g : \Lambda^2 T(M) \to \Lambda^2 T(M)$$

be the associated curvature operator. Take any symmetric function $\sigma : \mathbb{R}^{m(m-1)/2} \to \mathbb{R}$ and define $F_\sigma : S(\gamma) \to \mathbb{R}$ by

$$F_\sigma(g) = \int_M \sigma(R_g)\nu_g$$

where $\sigma(R_g) : M \to \mathbb{R}$ assigns to each point of M the σ-function of the eigenvalues of R_g.

An interesting special case occurs if $m \geqslant 3$. *Taking for* σ *the first elementary symmetric function* σ_1 *, then the variational density of* F_σ *becomes the total curvature function, and its critical points are just the Einstein metrics on* M. (Theorem of Hilbert (7); see also (3, 8, 10)).

Example 4. Let M,N be compact Riemannian manifolds and $\gamma : M \times N = W \to M$ be projection on the first factor. We can identify the sections of γ with the maps $\phi : M \to N$. For any $k \geqslant 1$ let $T_{(k)}(M) \to M$ denote the vector bundle of $k^{\underline{th}}$ order tangent vectors on M (see (11)); thus local sections of $T_{(k)}(M)$ have representations in charts of M of the form

$$\sum_{|\alpha| \leqslant k} a_\alpha \frac{\partial^{|\alpha|}}{\partial x^\alpha} .$$

A smooth map $\phi : M \to N$ induces a linear bundle map (the kth *differential of* ϕ).

$$\phi_{(k)} : T_{(k)}(M) \to T_{(k)}(N).$$

The given Riemannian metrics on M,N naturally induce Riemannian metrics on $T_{(k)}(M), T_{(k)}(N)$. Thus if $r_{(k)}$ denotes the fibre dimension of $T_{(k)}(M)$ and $\sigma : \mathbb{R}^{r(k)} \to \mathbb{R}$ is a symmetric function, then the function $x \to \sigma(\phi^{\star}_{(k)} \circ \phi_{(k)})(x)$, which assigns to each $x \in M$ the σ-function of the eigenvalues of the endomorphism $\phi^{\star}_{(k)} \circ \phi_{(k)}$, is a kth order variational density. As in Example 3, we choose $S(\gamma)$ to fit well with the functional $E_\sigma : S(\gamma) \to \mathbb{R}$ defined by

$$E_\sigma(\phi) = \int_M \sigma(\phi^{\star}_{(k)}\phi_{(k)})\ \nu_g.$$

Example 4 bis. In example 4 let us take $k = 1$ and $\sigma = \sigma_1/2$, where σ_1 is the first elementary symmetric function of m variables ($m = \dim M$). Then

$$\sigma(\phi^{\star}_{(1)} \circ \phi_{(1)})(x) = \frac{1}{2}|d\phi(x)|^2,$$

1/2 the Hilbert-Schmidt norm^2 of the differential $d\phi(x) : T_x(M) \to T_{\phi(x)}(N)$. Then $E_{\sigma_1/2}(\phi) = E(\phi)$ is the *energy of* ϕ. Its Euler-Lagrange equation is the semi-linear elliptic system

$$\mathrm{Trace}_g(\nabla d\phi) \equiv \mathrm{div}(d\phi) \equiv 0;$$

and the critical points of the energy are called *harmonic maps*.

In general the choice of $S(\gamma)$ is not a simple matter. In the special case of energy the fundamental regularity theorem (6,3.5)) suggests taking $S(\gamma) = \mathcal{L}^2_1(\gamma) \cap C^0(\gamma)$.

With that choice

(1) $S(\gamma)$ *is a smooth Banach manifold with natural complete Finsler structure*. (Details of that result have been supplied by F. E. Burstall, Warwick Thesis (in preparation).)

(2) $E : S(\gamma) \to \mathbb{R}$ *is a smooth function. Its critical points (in the differentiable sense) are just the harmonic maps.*

(3) *The components of* $S(\gamma)$ *are just the homotopy classes of maps* $M \to N$, *of the class* $\mathcal{L}^2_1 \cap C^0$.

Remark. In Examples 4 and 4 bis it is not difficult to replace the product bundle $\gamma : M \times N \to M$ by an arbitrary Riemannian fibre bundle $\gamma : W \to M$. A section ϕ of γ has differential which decomposes into horizontal and vertical parts : $d\phi = (d\phi)^H + (d\phi)^V$, following the vector bundle decomposition of T(W); and we can construct σ-densities of $(d\phi)^V$. In particular, the extremals of the energy functiona

$$E(\phi) = \frac{1}{2}\int_M |(d\phi)^V|^2 \nu_g$$

are the *harmonic sections of* γ . Their study has been undertaken by C. M. Wood at Warwick.

Example 5. Let $\gamma : M \times N = W \to M$ be as in Example 4, so that we again identify sections of γ with maps $\phi : M \to N$.

Let $\lambda_1 \geq \ldots \geq \lambda_r > 0$ be the positive eigenvalues of $\phi^*_{(1)} \circ \phi_{(1)}(x)$ at a given point; and set $\mu_i = \lambda_i^{1/2}$.

Then for any real number $p \geq 1$ we have the norm

$$\| d\phi(x) \|_{(p)} = (\Sigma\, \mu_i^p)^{1/p} ,$$

which for p = 2 reduces to the Hilbert-Schmidt norm of Example 4 bis. Then of course we can define

$$E_{(p)}(\phi) = \int_M \| d\phi(x) \|^p_{(p)} dx$$

and study its extrema. Regularity theory for such integral problems seems to be substantially more subtle and difficult than for the case p = 2. However, *the space*

$$S(\gamma) = \mathcal{L}^p_1 \cap C^o$$

is again a complete smooth Finsler manifold of maps.

Particularly important is the case p = dim M = m; then $E_{(x)}$ has conformal invariance properties. That case has been studied by K. Uhlenbeck, who has established some existence theorems, which are both promising for future developments, and cautionary against over-optimism concerning regularity. See (13, 14, 12).

Example 5 bis. Taking $p = \infty$ in Example 5, we obtain

$$\| d\phi(x) \|_{(\infty)} = \sup\{ |d\phi(x)u|/|u| \ : 0 \neq u \in T_x(M)\} ;$$

and we can therefore define

$$E_{(\infty)}(\phi) = \sup\{\| d\phi(x) \|_{(\infty)} \ : x \in M\}.$$

This is the *Dehnung* of Olivier. It has been studied in detail in the case of maps between Euclidean spheres; but certainly deserves general consideration, as well. See (15), and the references therein.

Example 6. Let K and L be finite dimensional vector spaces and set $A = \odot^{*} L$, the symmetric algebra of L. A symmetric bilinear map $\beta: K \times K \to L$ shall be considered as such a map $\beta : K \times K \to A$. For a symmetric function $\sigma . A \times A \to A$ we can define $\sigma(\beta) \in A$ as the σ-function of the eigenvalues of β (as elements of A).

Now given a map $\phi : M \to N$ as in Example 4 bis, its second fundamental form

$$\beta\phi = \nabla d\phi \in C(\odot^2 T^{*}(M) \otimes \phi^{-1}T(N));$$

see (6,§3). And we can apply the above construction to obtain the tensor field $\sigma(\beta\phi)$. In particular, taking $\sigma = \sigma_m$ and $N = \mathbb{R}$ we obtain the *Monge-Ampère operator* $\sigma_m(\beta\phi)$. Certain of these operators are in fact Euler-Lagrange operators associated with variational principles; e.g., look at the case m = 2. This general theory presents many enticing mysteries-about which we reluctantly make no further comment at this time.

REFERENCES

(1) M. F. Atiyah, Complex analytic connections in fibre bundles. Trans. Amer. Math. Soc. 85 (1957), 181-207.

(2) M. F. Atiyah, N. J. Hitchin, and I.M. Singer, Self-duality in four-dimensional Riemannian geometry. Proc. Roy. Soc. London A 362 (1978), 425-461.

(3) M. Berger, Quelques formules de variation pour une structure riemannienne. Ann. E.N.S. 3 (1970), 285-294.

(4) J. Eells and J. H. Sampson, Harmonic mappings of Riemannian manifolds. Ann. J. Math. 86 (1964), 109-160.

(5) J. Eells and J. H. Sampson, Variational theory in fibre bundles. Proc. U.S.-Japan Sem. Diff. Geometry. Kyoto, (1965), 22-33.

(6) J. Eells and L. Lemaire, A report on harmonic maps. Bull. London Math. Soc. 10 (1978), 1-68.

(7) D. Hilbert, Die Grundlagen der Physik, Nachr. Ges. Wiss. Göttingen (1915), 395-407; (1917), 53-76. And Math. Ann. 92 (1924), 1-32.

(8) Y. Mutō , Critical Riemannian metrics. Tensor N.S. 29 (1975), 125-133.

(9) H. K. Nickerson, On differential operators and connections. Trans. Amer. Math. Soc. 99 (1961), 509-539.

(10) E. M. Patterson, A class of critical Riemannian metrics. Univ. Aberdeen (1980).

(11) W. F. Pohl, Differential geometry of higher order. Topology 1 (1962), 169-211.

(12) K. Suominen, Quasiconformal maps in manifolds. Ann. Acad. Fenn. 1966. No. 393.

(13) K. Uhlenbeck, Minimal 2-spheres and tori in S^k. Preprint (1975).

(14) K. Uhlenbeck, Equivariant harmonic maps into spheres. Proc. Tulane Conf. Harmonic Maps. (1981).

(15) H. Hefter, Dehnungsuntersuchungen an Abbildungen von Sphären auf Sphären. Bonn (1980).

CURVATURES OF TUBES ABOUT SUBMANIFOLDS

L. Gheysens
Aspirant van het Belgisch N.F.W.O.

I. INTRODUCTION

Let (M,g) be an n-dimensional Riemannian manifold of class C^{ω} and let P be a q-dimensional submanifold of M. A (solid) tube of radius r about P is the set

$$T(P,r) = \{\exp_m(x) \mid m \in P,\ x \in T_m^{\perp}(P),\ \|x\| \leq r\},$$

where $T_m^{\perp}(P)$ denotes the normal space of P at the point m. For small r the set

$$P_r = \{p \in T(P,r) \mid d(p,P) = r\}$$

is a smooth hypersurface which we also call a tube.

In 1939 H. Weyl (14) computed the volume $V_P(r)$ of P_r in the case that M is the Euclidean space E^n or a sphere S^n. For E^n he obtained

$$V_p(r) = \frac{(\pi r^2)^{\frac{n-q}{2}}}{(\frac{n-q}{2})!} \sum_{c=0}^{[\frac{q}{2}]} \frac{k_{2c}(R^T)}{(n-q+2)(n-q+4)\ldots(n-q+2c)} r^{2c},$$

where R^T is the curvature tensor of P and the functions k_{2c} are the even integrated mean curvatures of P, which are intrinsic in the case that M is E^n, S^n (or the hyperbolic space H^n). In this way H. Weyl proved that the volume $V_P(r)$ does not depend on the embedding of P when the ambient space is E^n or S^n but only depend

M. Cahen et al. (eds.), Differential Geometry and Mathematical Physics, 159–166.

on the intrinsic geometry of P.

Recently A. Gray and L. Vanhecke (12) computed the first two nonzero terms in the power series expansion for the volume of a tube about a submanifold P of an arbitrary Riemannian manifold M. They obtained the following formula :

$$V_P^M(r) = \frac{(\pi r^2)^{\frac{n-q}{2}}}{(\frac{n-q}{2})!} \int_P \{1 + Ar^2 + O(r^4)\}(m)dm,$$

where $A = \frac{-1}{6(n-q+2)} (\tau - 3\tau^T + \sum_{a=1}^{q} \rho_{aa} + \sum_{a,b=1}^{q} R_{abab})$.

(R,ρ,τ are the curvature tensor, the Ricci tensor and the scalar curvature of M and τ^T is the scalar curvature of P; $\{e_a, a=1,\ldots,q\}$ is an orthonormal basis for the tangent space of P). For all the rank one symmetric spaces they computed the complete series and proved in this way that already for tubes about surfaces in the complex projective space the volume does depend on the embedding.

In order to obtain this result they computed a power series expansion for the volume form and worked out the integration of the coefficients. This technique was applied in different situations :

i) dim P = 0 ((7),(9),(10)). Then P_r is a geodesic sphere with radius r and one can use power series expansions in normal coordinates. It is interesting to see to what extent the geometry of the geodesic spheres determines the geometry of the ambient space ((3),(4)).

ii) dimP = 1 ((11)). For tubes about curves they used power series expansions in Fermi coordinates and proved

<u>Theorem 1</u>. *In E^n and in all the rank one symmetric spaces (model spaces) the volume of a small tube about a curve σ does not depend on the embedding of σ.*

iii) dim P = q, q < n ((12)). Here they used power series expansions in generalized Fermi coordinates.

To compute the volume element of a geodesic sphere there is another method, which uses the so-called recursion formula of

Ledger and provides an elegant way to compute the second fundamental form and the mean curvature of the geodesic spheres (see (2),(4),(13)). We generalized this method for tubes about curves (6). For tubes about arbitrary submanifolds we refer to (8), (12).

II. POWER SERIES EXPANSIONS FOR TUBES ABOUT CURVES

Let ∇ be the Riemannian connection and let R be the curvature operator of (M,g), i.e.

$$R_{XY}Z = \nabla_{[X,Y]}Z - \nabla_X\nabla_Y Z + \nabla_Y\nabla_X Z \ , \quad X,Y,Z \in \mathcal{H}(M).$$

Let $\sigma : [a,b] \to M$ be a unit speed curve in M and suppose that p is a point on the tube P_s such that γ is the geodesic of length s meeting σ orthogonally at $m = \sigma(0)$. We assume that γ is parametrized by its arc length s and that $\gamma(0) = m$. To describe the geometry of M in the neighborhood of σ we use Fermi coordinates and Fermi vector fields.

Definition 1. (11). Let $\{E_1,\ldots,E_n\}$ be an orthonormal frame field along a curve σ: $[a,b] \to M$ and let $m=\sigma(0)$ be a point on σ. Assume that $\sigma(t) = E_1\Big|_{\sigma(t)}$. The Fermi coordinates $(x_1,\ldots,x_n)$ of (M,σ) relative to $\{E_1,\ldots,E_n\}$ and m are given by

$$x_1(\exp_{\sigma(t)} \sum_{j=2}^{n} t_jE_j\Big|_{\sigma(t)}) = t,$$

$$x_i(\exp_{\sigma(t)} \sum_{j=2}^{n} t_jE_j\Big|_{\sigma(t)}) = t_i, \quad 2 \leqslant i \leqslant n.$$

The Fermi coordinates are defined in an open neighborhood $\mathcal{U}$ of σ.

Definition 2. (11). Let $(x_1,\ldots,x_n)$ be a Fermi coordinate system of (M,σ) relative to $(E_1,\ldots,E_n)$ and m. Then $X \in \mathcal{H}(\mathcal{U})$ is a Fermi vector field for (M,σ) provided that

$$X = \sum_{i=2}^{n} c_i \frac{\partial}{\partial x_i} \ ,$$

where the c_i's are constants.

We put $\frac{\partial}{\partial x_1} = A$. Then the curvature κ and the torsion $\perp$ of satisfy the relations :

$$(\nabla_A A)\Big|_{\sigma(t)} = \kappa(\sigma(t))U(\sigma(t)),$$

where U is a unit vector field along σ, everywhere perpendicular to σ and

$$(\nabla_A X)\Big|_{\sigma(t)} = \{-\kappa g(X,U)A + \perp X\}\Big|_{\sigma(t)},$$

for any Fermi vector field $X \in \mathcal{H}(U)$.

So we have that

$$\dot{\sigma}(t) = \frac{\partial}{\partial x_1}\Big|_{\sigma(t)} = E_1(t) = A\Big|_{\sigma(t)},$$

$$\frac{\partial}{\partial x_i}\Big|_{\sigma(t)} = E_i(t), \qquad i=2,\ldots,n,$$

and we take the basis $\{E_1,\ldots,E_n\}$ such that

$$\frac{\partial}{\partial x_2}\Big|_{\sigma(o)} = E_2(0) = \gamma'(0).$$

We denote by $\{e_1,\ldots,e_n\}$ the orthonormal frame along γ obtained by parallel translation of $\{E_i(0), i=1,\ldots,n\}$ along γ.

In order to generalize the Ledger technique for tubes about curves we need 4 fundamental observations (see (2),(6),(8)).

1. <u>The use of Jacobi vector fields along</u> γ

They are obtained by solving the Jacobi field equation

$$\nabla_{\gamma'}\nabla_{\gamma'}Y + R_{\gamma'Y}\gamma' = 0 \quad .$$

We choose the initial conditions as follows :

$$Y_1(0) = E_1(0)\ ,\ Y_1' = E_1'(0) = (\nabla_{\gamma'}E_1)(m)\ ,$$

$$Y_\alpha(0) = 0\ ,\ Y_\alpha'(0) = E_\alpha(0)\ ,\alpha = 3,\ldots,n.$$

Then the Jacobi vector fields along γ are given by

$$Y_1(s) = \frac{\partial}{\partial x_1}\Big|_{\gamma(s)}\ ,\ Y_2(s) = \frac{\partial}{\partial x_2}\Big|_{\gamma(s)}\ ,$$

$$Y_3(s) = s\,\frac{\partial}{\partial x_3}\Big|_{\gamma(s)},\ldots,\ Y_n(s) = s\,\frac{\partial}{\partial x_n}\Big|_{\gamma(s)}.$$

2. <u>The endomorphism</u> $B(s)$ <u>of the tangent space</u> $T_{\gamma(s)}M$

It is defined by

$$B(s)e_a = Y_a(s).$$

Then we have the following relations :

i) B satisfies the endomorphism-valued equation

$B'' + R \circ B = 0$, where R denotes the endomorphism of $T_{\gamma(s)}M$ given by $R(s)x = R_{\gamma' x}\gamma'$.

ii) When ω is the volume form of M and when we put

$$\theta(p) = \omega\left(\frac{\partial}{\partial x_1}, \ldots, \frac{\partial}{\partial x_n}\right)_p ,$$

then

$$\det B(s) = s^{n-2}\theta(p).$$

iii) By a classical formula for the derivative of a determinant we have that

$$\frac{(\det B(s))'}{\det B(s)} = \operatorname{trace}\,(B'B^{-1})\,(s).$$

3. <u>The shape operator</u> S <u>of the tube</u> P_s

It satisfies the relation

$$S = B'B^{-1}$$

and hence the mean curvature h of P_s is given by

$$h(p) = \operatorname{trace} S(s) = \frac{n-2}{s} + \frac{\theta'(s)}{\theta(s)} .$$

4. <u>The Ledger recursion formula</u>

Put $C = sB'B^{-1}$, then we have $C(p) = \sum_{k=0}^{\infty} \frac{s^k}{k!} C^{(k)}(0)$, where the coefficients are given by :

$$\forall n \in \mathbb{N}_o : (n-1)C^{(n)}(0) = - n(n-1)R^{(n-2)}(0) - \sum_{k=0}^{n} \binom{n}{k} C^{(k)}(0)\, C^{(n-k)}(0).$$

In particular we have for tubes about curves that

$$C(0) = \left[\begin{array}{c|ccc} 0 & 0 & \ldots & 0 \\ \hline 0 & & & \\ \vdots & & I & \\ 0 & & & \end{array}\right],$$

$$C'(0) = \left[\begin{array}{c|ccc} -\kappa(m)g(U,E_2)(m) & 0 & \ldots & 0 \\ \hline 0 & & & \\ \vdots & & 0 & \\ 0 & & & \end{array}\right]$$

III. RESULTS FOR TUBES ABOUT CURVES

The preceding expressions imply

Theorem 2. *The endomorphism* C, *the shape operator* S, *the mean curvature* h, *the volume element (and the volume of a small tube* P_s *about a curve* σ *in an* n*-dimensional Riemannian manifold do not depend on the torsion of* σ.

Using the Gauss equation

$$R^T(X,Y,Z,W) = R(X,Y,Z,W) + g(SX,Z)g(SY,W) - g(SX,W)g(SY,Z),$$

where $X,Y,Z,W \in \mathcal{K}(P_s)$, we obtain

Theorem 3. *The curvature operator* R^T , *the Ricci tensor* ρ^T *and the scalar curvature* τ^T *of a small tube* P_s *about a curve* σ *in an* n*-dimensional Riemannian manifold do not depend on the torsion of* σ.

In (5) we investigate the second fundamental form, the total mean curvature $H_\sigma(s) = \int_{P_s} h(p)dp$, the total scalar curvature $T_\sigma(s) = \int_{P_s} \tau^T(p)dp$, the total Lipschitz-Killing curvature $L(P_s) = (-1)^{n-1} \int_{P_s} (\det S)(p)(dp)$, the Weyl tensor C^T and several other functions for tubes about curves. For all the model spaces we give the complete power series expansions for these functions. In this way we prove

Theorem 4. *Let* M *be an* n*-dimensional Riemannian manifold* $(n \geq 3)$. *Then* M *is a space of constant curvature if and only if all sufficiently small tubes about an arbitrary geodesic are quasi-umbilical hypersurfaces, where the special direction is determined by the parallel displaced tangent vector to the curve. If in addition the principal curvature for that particular direction is equal to zero, then* M *is locally flat.*

(A similar result is proved for the Ricci tensor).

Theorem 5. *Let* M *be an* n*-dimensional Riemannian manifold* $(n \geq 4)$. *Then* M *is a space of constant curvature if and only if all sufficiently small tubes about an arbitrary geodesic are conformally* *flat hypersurfaces.*

Theorem 6. *Let* M *be an* n*-dimensional Riemannian manifold* (n > 3) *with adapted holonomy group (as in the model space) and with the property that for all small* s *and all sufficiently short geodesics* σ M *has the same total scalar curvature functions* $T_\sigma(s)$ *as in the model space. Then* M *is locally isometric to that model space.*

Theorem 7. *Let* σ *be a curve in Euclidean space or in a rank one symmetric space. Then the total mean curvature* $H_\sigma(s)$, *the total scalar curvature* $T_\sigma(s)$ *and the total Lipschitz-Killing curvature* $L(P_s)$ *of a small tube* P_s *about* σ *do not depend on the embedding of* σ.

IV. TUBES ABOUT ARBITRARY SUBMANIFOLDS

In (12) the Ledger technique is generalized for tubes about arbitrary submanifolds. Also in this situation the torsion of the submanifold (corresponding to the normal connection) disappears in most of the formulas. Combining the formula for the volume of tubes about an arbitrary submanifold with the Steiner formulas for parallel hypersurfaces (1) it is possible to prove

Theorem 8 (12). *Let* P_1 *and* P_2 *be isometric submanifolds of* E^n, *the sphere* $S^n(\lambda)$ $(\lambda > 0)$ *or the hyperbolic space* $H^n(\lambda)$ $(\lambda < 0)$ *and let* P_{1r} *and* P_{2r} *be tubes of small radius* r *about* P_1 *and* P_2. *Then all the integrated mean curvatures of* P_{1r} *are the same as those of* P_{2r}.

REFERENCES

(1) E. ABBENA, A. GRAY & L. VANHECKE, Steiner's formula for the volume of a parallel hypersurface in a Riemannian manifold, to appear in Ann. Scuola Norm. Pisa Cl. Sci..

(2) A.L. BESSE, Manifolds all of whose geodesics are closed, Ergebnisse der Mathematik: 1978, vol. 93, Springer-Verlag.

(3) B.-Y. CHEN & L. VANHECKE, Total curvatures of geodesic spheres: 1979, Arch. Math. (Basel) 32, pp. 404-411.

(4) B.-Y. CHEN & L. VANHECKE, Differential geometry of geodesic

spheres, to appear in J. Reine Angew. Math.

(5) L. GHEYSENS, doctoral dissertation, in preparation.

(6) L. GHEYSENS & L. VANHECKE, Total scalar curvature of tubes about curves, to appear in Math. Nachr..

(7) A. GRAY, The volume of a small geodesic ball in a Riemannian manifold: 1973, Michigan Math. J. 20, pp. 329-344.

(8) A. GRAY, Comparison theorems for the volumes of tubes, to appear.

(9) A. GRAY & L. VANHECKE, Riemannian geometry as determined by the volumes of small geodesic balls: 1979, Acta Math. 142, pp. 157-198.

(10) A. GRAY & L. VANHECKE, Oppervlakten van geodetische cirkels op oppervlakken: 1980, Med. Konink. Acad. Wetensch. Lett. Schone Kunst. België Kl. Wetensch. 42.

(11) A. GRAY & L. VANHECKE, The volumes of tubes about curves in a Riemannian manifold, to appear in Proc. Lond. Math. Soc..

(12) A. GRAY & L. VANHECKE, The volumes of tubes in a Riemannian manifold, to appear.

(13) H. RUSE, A.G. WALKER & T.J. WILLMORE: 1961, Harmonic Spaces, Cremonese , Roma.

(14) H. WEYL, On the volumes of tubes: 1939, Amer. J. Math. 61, pp. 461-472.

ALMOST SYMMETRIC SPACES

E.A. Ruh

University of Bonn

I. INTRODUCTION

Symmetric spaces, in particular Lie groups, play a central role in mathematics as well as in physics. It is therefore not surprising that qualitative versions of these objects should be of interest too. In fact they are. In differential geometry the study of these objects is conducted in the guise of pinching theorems, in physics it is the study of soft group manifolds introduced in (7) for non-internal gauge theories.

For the purpose of this lecture, let us restrict attention to riemannian symmetric spaces $\bar{M}$ as models and call a manifold M almost symmetric (of type $\bar{M}$) if each tangent space T_xM of M carries the structure of the tangent space of $\bar{M}$. As in the case of an almost complex structure the almost symmetric structure can be defined in terms of a reduction of the structure group of the principal bundle of frames on M. Again, as in the case of an almost complex structure the deviation from the model is measured in terms of a tensor, in our case the Cartan curvature tensor Ω.

This lecture is essentially a report on results obtained together with Min-Oo (5), (6). We prove that if the deviation from the model is small, then an almost symmetric structure can be

M. Cahen et al. (eds.), Differential Geometry and Mathematical Physics, 167–172.

deformed into a symmetric structure.

H.E. Rauch (8), (9) was the first mathematician to consider almost symmetric spaces in connection with comparison theorems. Rauch's first model was $\bar{M} = S^n$, the standard sphere.
Since the tangent space of a sphere carries no structure in addition to the metric, any riemannian manifold is almost a sphere in the above sense.

To obtain results the "almost" has to be qualified. Rauch's result is that $\|\Omega\| = \|R - R_1\|$ small enough, where R is the riemannian curvature of M and R_1 is the curvature of S^n, implies that the universal cover $\hat{M}$ of M is homeomorphic to S^n. Since 1950 much work has been done to weaken the hypothesis and to strengthen the assertion of this theorem. For details of this story we refer to (1).

II. THE RESULTS

Let G denote a compact semi-simple Lie group, G its Lie algebra and $\bar{\omega} : TG \to G$ the Maurer-Cartan form. An almost group structure on a differentiable manifold P is by definition simply a parallelization $\omega : TP \to G$ of P, where ω is a vector space isomorphism at every tangent space. The integrability condition is the validity of the Maurer-Cartan equation, therefore the Cartan curvature $\Omega = d\omega + [\omega, \omega]$ with $[\ ,\ ]$ the Lie bracket in G measures the extent of failure of integrability. Let $\|\Omega\|$ denote the maximum norm of Ω in terms of the natural metrics defined on G and TP respectively by the Cartan-Killing form of G. The first result is as follows.

<u>Theorem 1</u>. *Let G be a compact semi-simple Lie algebra, $\omega : TP \to G$ a parallelization of the compact manifold* P, *and $\Omega = d\omega + [\omega, \omega]$ the Cartan curvature of ω.*
There exists a positive constant A *depending only on G such that $\|\Omega\| < A$ implies that* P *is diffeomorphic to a quotient $\Gamma\backslash G$, where* G *is the simply connected Lie group with Lie algebra G and $\Gamma \subset G$ is a finite subgroup.*

To formulate the next result let $\bar{M}$ denote an irreductible riemannian symmetric space, M a compact almost symmetric space of type $\bar{M} = G/K$. This means by definition that the bundle of frames on M is reduced to a subbundle P with structure group K. In addition, we assume that $\eta : TP \to \mathcal{K}$, where $\mathcal{K}$ is the Lie algebra of K, is a connection on P. We define

$$\omega : \eta + \Theta : TP \to \mathcal{K} \oplus \mathbb{R}^n = \mathcal{G}$$

where $\Theta : T_uP \to \mathbb{R}^n$ is defined by $X \to u^{-1}\pi(X)$, $\pi : P \to M$ the projection and $u : \mathbb{R}^n \to T_{\pi(u)}M$ the linear isomorphism defined by the frame $u \in P$. With these definitions ω is a Cartan connection and defines a parallelization of P. $\Omega = d\omega + [\omega, \omega]$, where $[\ ,\ ]$ is the Lie bracket in $\mathcal{G}$, is the Cartan curvature of ω. Our result is as follows.

Theorem 2. *Let* $\bar{M} = G/K$ *denote an irreducible riemannian symmetric space. If* $\bar{M}$ *is of non-compact type assume in addition rank* $\bar{M} > 1$ *and* $\dim \bar{M} > 6$. Ω *the Cartan curvature of a compact manifold* M. *There exists a positive constant* A *depending on* $\bar{M}$ *and an upper bound of the diameter of* M *such that* $\|\Omega\| < A$ *implies that* M *is diffeomorphic to a quotient* $\Gamma\backslash\bar{M}$, *where* Γ *is a discrete subgroup of the isometry group* G *of* $\bar{M}$.

As mentioned above the metric on P and M respectively is defined by the Cartan-Killing form. In case $\bar{M}$ is of compact type Theorem 2 is a corollary of Theorem 1, we simply apply Theorem 1 to the principal bundle P and pass to quotients by factoring by K. In case $\bar{M} = S^n$, $\Omega = R - R_1$, where R is the riemannian curvature of M and R_1 is a curvature of S^n (after the obvious identifications). Thus, the above result is a generalization of the differentiable sphere pinching theorem.

We finish this section with a conjecture. It is a generalization of Theorem 1 and the result of Gromov (3), compare also (2), on almost flat manifolds. In fact even in these special cases it would give more precise results. The motivation for the assumptions of this conjecture is, as mentioned in the introduction, the wish to consider a qualitative version of the concept of Lie group.

The idea is to soften the rigid concept of left invariant vector fields.

Conjecture. Let T_o denote the Lie bracket of a compact Lie group, T and R respectively torsion and curvature of a metric connection on a compact riemannian manifold M with diameter d. There exists $A > 0$ depending only on $n = \dim M$ such that $(\|T - T_o\|^2 + \|\nabla T\| + \|R\|)d^2 < A$ implies that M is diffeomorphic to $\Gamma\backslash G$, where G is a Lie group, Γ satisfies the exact sequence $1 \to L \to \Gamma \to H \to 1$, L is a lattice in G and H is a finite subgroup of the automorphism group of G. In case $T_o = 0$ the manifold is called almost flat and G would have to be a nilpotent Lie group. If we assume in addition $T = 0$ and $R = 0$ then the conjecture reduce to the Bieberbach theorem on euclidian space forms.

III. SKETCH OF PROOF

To give a rough idea of the proof it suffices to restrict attention to Theorem 1. The basic reason why the proof works is the vanishing of the second cohomology $H^2(G,\mathbb{R})$ of a compact semi-simple Lie group with real coefficients. In case of Theorem 2 and $\bar{M}$ of non-compact type the relevant cohomology was introduced by Matsushima and Murakami (4) as a generalization of the Eichler cohomology.

The main work in the proof is to establish the existence of a Maurer-Cartan form $\bar{\omega} : TP \to \mathcal{G}$. We apply the method of Newton-Kolmogorov-Moser and solve a linearized deformation equation $d'\alpha = -\Omega$ approximately and iterate.
To obtain $\bar{\omega} = \omega + \alpha$ with vanishing curvature we must solve the equation

$$d\alpha + [\omega, \alpha] + [\alpha, \omega] + [\alpha, \alpha] = -\Omega.$$

To prove that $\bar{\omega}$ is again a parallelization and therefore a Maurer-Cartan form, we establish the estimate $\|\alpha\| < 1$.
Let

$$d'\alpha = -\Omega$$

denote the linearized equation. We introduce the Laplacian

$\Delta' = d'\delta' + \delta' d'$, where δ' is the adjoint of d' and define

$$\alpha = \delta'\beta ,$$

where β is the unique solution of the potential equation

$$\Delta'\beta = -\Omega.$$

Existence and uniqueness of β follows from the strict positivity of Δ'. We prove this by computing the Bochner formula for Δ':

$$\langle \Delta'\beta,\beta\rangle - \frac{1}{2}\Delta|\beta|^2 - |\tilde{D}\beta|^2 > \frac{1}{4}|\beta|^2 + O(\|\Omega\| \|\beta\|^2),$$

where Δ denotes the Laplacian on functions and $\tilde{D}$ is some operator that occurs in the computation. In case $P = G$ a compact semi-simple Lie group, we have $\Delta' = \Delta$, the Laplacian on 2-forms with values in $\mathbb{R}^N$, $N = \dim P$, and the above estimate proves the vanishing of $H^2(G,\mathbb{R}^N)$. The basic reasons why $\alpha = \delta'\beta$ solves the equation $d'\alpha = -\Omega$ well enough for the iteration to converge are the Bianchi equation $d'\Omega = 0$ and the fact that $d' \circ d'$ is an operator with norm $\|\Omega\|$. The main tools for the proof are a generalized maximum principle for elliptic operators and Sobolev inequalities.

REFERENCES

(1) Yu. D. Burago and V. A. Zalgaller, Convex sets in riemannian spaces of non-negative curvature, Russian Math. Surveys 32:3 (1977), 1 - 57.

(2) P. Buser and H. Karcher, Gromov's almost flat manifolds, Astérisques, Soc. Math. France, to appear.

(3) M. Gromov, Almost flat manifolds,J. Differential Geometry 13 (1978) 231-241.

(4) Y. Matsushima and S. Murakami, On vector bundle valued harmonic forms and automorphic forms on symmetric riemannian manifolds, Ann. Math. 78 (1963), 365 - 416.

(5) Min-Oo and E.A. Ruh, Comparison theorems for compact symmetric spaces, Ann. scient. Ec. Norm. Sup. 12 (1979), 335 - 353.

(6) Min-Oo and E.A. Ruh, Vanishing theorems and almost symmetric spaces of non-compact type, Math. Ann., to appear.

(7) Y. Ne'eman and T. Regee, Gravity and supergravity as gauge theories on a group manifold. Physics letters 74 B(1978), 54 - 56.

(8) H. E. Rauch, A contribution to riemannian geometry in the large, Ann. of Math. 54, 38 - 55.

(9) H. E. Rauch, Geodesics, symmetric spaces, and differential geometry in the large, Comment. Math. Helv. 27, (1953), 294 - 320.

MINIMAL SURFACES AND WEYL GROUPS

B. Smyth
University of Notre Dame, Indiana.

This lecture is about the existence of periodic minimal surfaces.

The first technique, originating with Schwarz (6), for the construction of such surfaces in $\mathbb{R}^n$ begins with a polygon P having the property:

(*) the group generated by the reflexions in the edges of P is a discrete uniform subgroup of the group of motions of $\mathbb{R}^n$.

P is then spanned by a solution of the Plateau problem and by repeated reflexion of this solution across boundaries (Schwarz reflexion principle) one obtains a periodic minimal surface. The surfaces obtained in this way will be called *Schwarz surfaces*.

The second technique is due to Weierstrass (8) and arose from his interest in describing the Schwarz surfaces by equations. The key ideas are the Weierstrass representation and the Jacobi inversion theorem. The surfaces constructed are described as the real zeroes of theta functions. In the same paper Weierstrass conjectures that every flat 3-torus contains a compact imbedded minimal surface.

Nature may also lend a hand in the construction of such surfaces. Donnay-Pawson (1) reports on the appearance of what may

M. Cahen et al. (eds.), Differential Geometry and Mathematical Physics, 173–175.

be a triply-periodic minimal surface in crystalline biology. Periodicity is evident but minimality cannot be judged from a photograph; negative curvature this example clearly has.

In (4) both the Schwartz construction is studied as well as the geometric and analytic properties of the Schwarz surfaces. This gives an indirect method of constructing new periodic minimal surfaces from Schwarz surfaces (Theorem 3 below). An important element of that paper is the formula of Chevalley and Weil (9).

The main results of (4) are :

Theorem 1. P *satisfies* (*) *if and only if* P *is a root polygon (i.e. edges are integer multiples of the roots of some root system)*

Remark. The case P = quadrilateral in $\mathbb{R}^3$ was done by Schoenflies (5) but his method cannot handle more edges. So even for n = 3 the result is new.

Theorem 2. *Any solution of the Plateau problem for a polygon* P *with*

(i) n+1 *edges*

(ii) $\langle e_i, e_j \rangle \leqslant 0$ *for all pairs of edges*

(iii) *vertex angles of form* π/k *(*k *integer)*

is regular in the interior and at the boundary.

Remark . Theorems 1 and 2 guarantee the existence of Schwarz surfaces with no singularities.

Theorem 3. *Let* $f : X \to T^n$ *be a Schwarz surface obtained from a root polygon* P *satisfying* Theorem 2. *Let* $df = (df_1, \ldots, df_n)$ *and let* $dg = (dg_1, \ldots, dg_n)$ *denote the harmonic differentials conjugate to* df. *Then* $\int_X dg$ *defines another minimal immersion of* X *into another* n-*torus.*

In any space form, the characterisation of those geodesic polygons satisfying (*) is an attractive problem. Already in the spherical case important conclusions were drawn by Lawson (3). It looks as though the hyperbolic Coxeter groups can be used to provide such a characterisation in hyperbolic space c.f. Lanner (2) and Vinberg (7).

REFERENCES

(1) Donnay-Pawson, X-ray Diffraction Studies of Echinoderm Plates, Science, 166 (1969), 1147-1150.

(2) Lanner, On complexes and transitive groups of automorphisms, Lunds Univ. Math. Sem., 11 (1950).

(3) Lawson, Complete minimal surfaces in S^3, Ann of Math., 92(1970), 335-374.

(4) Nagano-Smyth, Periodic minimal surfaces and Weyl groups, Acta Math. 145 (1980), 1-27.

(5) Schoenflies, Sur les surfaces minima limitées par quatre arêtes d'un quadrilatère gauche, C.R. Acad. Sci. Paris 112 (1897), 478-480.

(6) Schwarz, Gesammelte Mathematische Abhandlungen Band 1, Springer, Berlin, 1890.

(7) Vinberg, Some arithmetical discrete groups in Lobaĉevskiî spaces, Bombay Colloquim 1973 on "Discrete subgroups of Lie groups", Oxford University Press, 1975.

(8) Weierstrass, Werke, Band III.

(9) Weil, Über Matrizenringer auf Riemannschen Flächen und den Riemann-Rochschen Satz, Abh. Math. Sem. Hamburg 11 (1936), 110-115.

SOME ASPECTS OF HARMONIC MAPS FROM A SURFACE TO COMPLEX PROJECTIVE SPACE

J. C. Wood

University of Leeds, Leeds LS2 9JT, U.K.

0. INTRODUCTION

Let M and N be smooth compact Riemannian surfaces without boundary. If N has negative curvature, or more generally, if $\pi_2(N) = 0$, we have general existence theorems for harmonic maps (8, 18, 22, 23). The first interesting case where we have no general existence theorem is when N is the two-sphere S^2. In §1 we discuss some results in this case and see how they generalize when $S^2 = \mathbb{C}\mathbb{P}^1$ is replaced by the n-sphere S^n or complex projective n-space $\mathbb{C}\mathbb{P}^n$. In §2 we outline a development of the method introduced by Din-Zakrewski (4), Buns and Glaser-Stora (13) for manufacturing certain harmonic maps into $\mathbb{C}\mathbb{P}^n$ from holomorphic ones. This method classifies all harmonic maps $S^2 \to \mathbb{C}\mathbb{P}^n$ and all harmonic maps from the torus to $\mathbb{C}\mathbb{P}^n$ of non-zero degree. Full details will appear in a paper by J. Eells and the author (11).

I. HARMONIC MAPS INTO SPHERES AND COMPLEX PROJECTIVE SPACES

1. <u>The fundamental questions in harmonic map theory</u>. Let M and N be smooth (= C^∞) Riemannian manifolds without boundary of any dimensions. For simplicity of exposition we assume these are compact. A *harmonic map* is a smooth map $\phi : M \to N$ which is a

M. Cahen et al. (eds.), Differential Geometry and Mathematical Physics, 177–186.

critical point of the *energy functional*

$$E(\phi) = \frac{1}{2}\int_M \| d\phi_x \|^2 v(x)$$

where $\| d\phi_x \|$ is the Hilbert-Schmidt norm of the linear map $d\phi_x : T_x M \to T_{\phi(x)} N$ and $v(x)$ is the volume element at $x \in M$. A smooth map is harmonic if and only if it satisfies the *harmonic* equation $\operatorname{div} d\phi = 0$ where div is the divergence operator on the bundle $T^* M \otimes \phi^{-1} TN$ induced from the metrics and Levi-Civita connections on M and N. For general details of harmonic maps see (8, 6, 26), for a more general class of variational problems see (5). The fundamental questions in harmonic map theory are : given a homotopy class h of maps from M to N, (i) *existence*: does there exist at least one harmonic map in h?, (ii) *uniqueness:* does there exist at most one harmonic map in h?, (iii) *classification :* can we describe all harmonic maps in h? The first main results in the theory are that we always have *existence* if N has non-positive Riemannian sectional curvature (8) and we further have *uniqueness*, except in some trivial cases, if N has strictly negative Riemannian sectional curvatures (one can be slightly more general see (15, 25)). Using such results we can employ properties of harmonic maps to study the fundamental group of N (27).

2. Now suppose that M is a surface. Then harmonicity depends only on the conformal structure of M (8). If $\pi_2(N) = 0$ we have *existence* (18, 23, 25) (but not, in general, *uniqueness*). If M is oriented, harmonicity depends only on the complex structure induced by the metric of M, we may thus refer to harmonic maps from a *Riemannian surface* M to a Riemannian manifold N.

3. Maps from a surface to the 2-sphere. We now discuss the first cases not covered by the existence and uniqueness theorems. Let M_p denote a closed (i.e. compact without boundary) Riemann surface of genus p. Recall that any $\pm$ holomorphic (i.e. holomorphic or anti-holomorphic) map between Kähler manifolds is harmonic (8) and is of absolute minimum energy within its homotopy class (20).

Conversely we have a

4. Holomorphicity Theorem.

(i) *Any harmonic map* $M_p \to S^2$ *of degree* $\geq p$ *is holomorphic* (9, 10, 11). (Here S^2 can have any metric). It follows that

(ii) *all harmonic maps* $S^2 \to S^2$ *are* $\pm$ *holomorphic,*

(iii) *there exists no harmonic map from a 2-torus* T^2 *to a 2-sphere* S^2 *of degree one whatever metrics these are given.* In contrast Lemaire shows (19):

(iv) *for any* d, $0 \leq d \leq p-1$, *there exists a Riemann surface* M_p *and a harmonic non-holomorphic map* $M_p \to S^2$ *of degree* d. (Here S^2 is given a sufficiently symmetrical metric). For related non-orientable cases see (7)).

We can generalize these results in two ways. Firstly, recall that a map $\phi : M \to N$ is said to be *(weakly) conformal* if there exists a smooth function $\lambda : M \to (0,\infty)$ such that $\| d\phi_x(X) \| = \lambda(x) \|X\|$ for all $x \in M$, $X \in T_xM$. Any $\pm$ holomorphic map from a Riemann surface M to an almost Hermitian manifold N is weakly conformal; if N is a Riemann surface the converse is also true. A non-constant weakly conformal map from a Riemann surface to a Kähler manifold is harmonic if and only if it is a *minimal branched immersion* in the sense of (14). Let z denote a complex coordinate on M. In terms of complex derivatives $\partial_z^c\phi = d\phi(\frac{\partial}{\partial z})$, $\partial_{\bar{z}}^c\phi \in T^cN$ a map $M \to N$ is weakly conformal if and only if, on each coordinate open set, the Hermitian inner product in T^cN, $\langle \partial_z^c\phi , \partial_{\bar{z}}^c\phi \rangle$, is identically zero. Theorem 4. (ii) generalizes to the theorem (26):

(v) *any harmonic map from* S^2 *to an arbitrary Riemannian manifold is weakly conformal.*

5. Maps from a surface to the n-sphere. Let S^n be the n-sphere with its standard metric. Say that a map $\phi : M \to S^n$ from a Riemann surface is isotropic or satisfies the Calabi identities (Calabi's terminology (3) : pseudoholomorphic) if

$$\langle (\partial_z)^\alpha\phi, (\partial_{\bar{z}})^\beta\phi \rangle \equiv 0 \text{ for all } \alpha \geq 0,\ \beta \geq 0,\ \alpha + \beta \geq 1 \tag{1}$$

Here ϕ denotes the composition of ϕ with the composite inclusion map $i : S^n \hookrightarrow \mathbb{R}^{n+1} \hookrightarrow \mathbb{C}^{n+1}$ and $<,>$ denotes Hermitian inner product on $\mathbb{C}^{n+1}$. Note that (6) makes sense also for maps ϕ from a surface into real projective n-space $\mathbb{R}\mathbb{P}^n$ by taking local lifts $\phi_U : U \to S^n$ and setting $\phi = i \circ \phi_U$ on U. Note that (1) is automatic for $\alpha + \beta = 1$. For $\alpha + \beta = 2$ it is the condition of weak conformality. In terms of the connection D^N on $\phi^{-1}T^cS^n$ (or $\phi^{-1}T^c\,\mathbb{R}\mathbb{P}^n$), (1), is equivalent to

$$<(D^N_z)^\alpha\phi,\ (D^N_{\bar{z}})^\beta\phi> \equiv 0 \text{ for all } \alpha \geqslant 1,\ \beta \geqslant 1. \qquad (2)$$

We have a generalization of I.4. (ii):

6. Isotropy Theorem. *Any harmonic map* $\phi : S^2 \to S^n$ *is isotropic i.e. satisfies* (1). This result does not seem to generalize when S^2 is replaced by a closed Riemann surface of higher genus. Indeed, it follows from Itoh and Kenmotsu (see (16)) that a weakly conformal harmonic map $T^2 \to S^n$ is never isotropic.

7. Maps from a surface to complex projective n-space. Let $\mathbb{C}\mathbb{P}^n$ denote complex projective n-space with its Fubini-Study metric of constant holomorphic sectional curvature 2. Homotopy classes of maps $\phi : M_p \to \mathbb{C}\mathbb{P}^n$ from a closed Riemann surface of genus p are classified by *degree* = degree of the induced map on second cohomology. We have generalizations of 3. (i) :

8. Holomorphicity Theorems.

(i) *All harmonic maps* $M_p \to \mathbb{P}^n$ *of degree* $\geqslant p$ *such that, for all* $x \in M_m$, $d\phi(T_xM) \subseteq$ *complex* 1-*dimensional subspace of* $T_{\phi(x)}N$, *or* ϕ *nearly enough satisfies this condition, then* ϕ *is holomorphic* (28).

(ii) *All harmonic maps* $M_p \to \mathbb{P}^n$ *of degree* $\geqslant 2p-1$ *such that the* (0,1) *energy*

$$E''(\phi) \equiv \int_M \|\bar{\partial}\phi_x\|^2 v(x) < 2\pi\{d(\phi) - (2p-2)\}$$

are holomorphic. See also (29, 30), and for theorems asserting holomorphicity of minimum energy maps, (10, 12, 24).

If N is a Kähler manifold we shall say that a smooth map $\phi : M \to N$ from a Riemann surface M is (complex) isotropic or satisfies the (*complex*) *Calabi identities* (earlier terminology *strongly pseudoholomorphic*) if, on each coordinate open set,

$$<(D'_z)^{\alpha}\phi, (D'_{\bar{z}})^{\beta}\phi> \equiv 0 \text{ for all } \alpha \geq 1, \quad \beta \geq 1. \qquad (3)$$

Here D' denotes the connection and <,> the Hermitian inner product on $\phi^{-1}T^{1,0}N$. For $\alpha + \beta = 2$ this condition is that of weak conformality. A smooth map between surfaces satisfies (3) if and only if it is $\pm$ holomorphic. The result 4.(i) generalizes in the cases p = 0 and 1 as follows :

9. Complex isotropy Theorem

(i) *Any harmonic map* $\phi : S^2 \to \mathbb{C}P^n$ is complex isotropic (11, 13).

(ii) *Any harmonic map* $\phi : T^2 \to \mathbb{C}P^n$ *from a 2-torus of non-zero degree is complex isotropic* (11). We have only partial generalizations to higher genus (11). (Note, S^2 and T^2 may have any metrics in these theorems.

10. Using the method of §2 to manufacture isotropic harmonic maps, we have, in contrast to 4.(i) : *there exist harmonic non-holomorphic maps* $M_p \to \mathbb{C}P^n$ of (i) *all degrees* (≥ 0) *if* p = 0 *or* 1, (ii) *degree* 0 *and all degrees* $\geq p + 1$ *if* $p \geq 2$ (11).

II. MANUFACTURING HARMONIC MAPS INTO COMPLEX PROJECTIVE SPACE

1. Let M be a Riemann surface. Let $\mathbb{C}P^n$ denote complex projective n-space with its Fubini-Study metric of constant holomorphic sectional curvature 2; let $\pi : \mathbb{C}^{n+1} \setminus 0 \to \mathbb{C}P^n$ denote the standard projection. Let $f : M \to \mathbb{C}P^n$ be a holomorphic map which is *full* i.e. its image lies in no projective subspace. We outline how to manufacture a harmonic map $\phi : M \to \mathbb{C}P^n$ by a reformulation and generalization of the method of Din-Zakrewski (4), Glaser-Stora (13) and Buns.

2. Let $G_{\alpha}(\mathbb{C}^{n+1})$ denote the Grassmannian of complex α-dimensional subspaces of $\mathbb{C}^{n+1}$. We define the *associated curves* $f_{\alpha} : M \to G_{\alpha+1}(\mathbb{C}^{n+1})$ ($\alpha = 0,1,\ldots,n$) of f as follows : Let $x \in M$, let $f_U : U \to \mathbb{C}^{n+1} \setminus 0$ be a *lift* of f over a coordinate open set U

of M which contains x (i.e. $f = \pi \circ f_U$). Let z denote a local complex coordinate on U. If the derivatives $f_U, \partial_z f_U, \ldots, \partial^{\alpha}_z f_U$ are linearly independent at x set $f_{\alpha}(x)$ = $(\alpha + 1)$ -dimensional subspace spanned by those derivatives. Then $f_{\alpha}(x)$ is well-defined independently of the chosen coordinate, and due to holomorphicity may be extended to points $x \in M$ where the derivatives are linearly dependent (see (31)) . Note that $f_o = f$; it is convenient to set f_{-1} = zero map. Using $\partial_{\bar z}$ - derivatives in place of ∂_z - derivatives we can define in a similar way the associated curves of an antiholomorphic map.

3. Now let $f : M \to \mathbb{C}\mathbb{P}^n$ be a full holomorphic map and consider the map $g = f^{\perp}_{n-1} : M \overset{f_{n-1}}{\to} G_n(\mathbb{C}^{n+1}) \overset{\perp}{\to} G_1(\mathbb{C}^{n+1}) = \mathbb{C}\mathbb{P}^n$ where $\perp$ assigns to a subspace its orthogonal complement. Then g is full and antiholomorphic and the pair (f,g) satisfies the orthogonality relations

$$f_{\alpha} \perp g_{\beta} \text{ for all } \alpha,\beta \geqslant 0, \quad \alpha + \beta \leqslant n - 1 . \qquad (4)$$

In terms of local lifts f_U and g_U of f and g this reads

$$\langle \partial^{\alpha}_z f_U, \partial^{\beta}_{\bar z} g_U \rangle \equiv 0 \text{ for all } \alpha,\beta \geqslant 0, \ \alpha + \beta \leqslant n - 1$$

where <,> denotes the Hermitian inner product on $\mathbb{C}^{n+1}$. Now for any $r \in \{0,1,\ldots,n\}$, $s = n-r$, the subspaces $f_{r-1}(x)$ and $g_{s-1}(x)$ are orthogonal hence the assignment

$$\phi(x) = (f_{r-1}(x) \oplus g_{s-1}(x))^{\perp} \quad , \quad x \in M \qquad (5)$$

defines a map $\phi : M \to \mathbb{C}\mathbb{P}^n$. Note that if r = 0, $\phi = g_{n-1}^{\perp} = f$ is holomorphic and if r = n, $\phi = f_{n-1}^{\perp} = g$ is antiholomorphic.

The map ϕ *is harmonic.* To see this note that it is the composition of the map $\psi = (f_{r-1}, g_{s-1}) : M \to \mathcal{H}_{r,s} = \{(V,W) \in G_r(\mathbb{C}^{n+1}) \times G_s(\mathbb{C}^{n+1}) : V \perp W\}$ and the map $\pi : \mathcal{H}_{r,s} \to \mathbb{C}\mathbb{P}^n$ defined by $\pi(V,W) = (V + W)^{\perp}$. Now π is a Riemannian submersion (21) and ψ is harmonic (since its components are $\pm$ holomorphic) and, because of the orthogonality relations, horizontal with respect to π, i.e. $d\psi(T_x M)$ lies in the horizontal subspace of $T_{\psi(x)}\mathcal{H}_{r,s}$ for each $x \in M$. It is easily seen (11) that the composition of such maps is harmonic. It can be further shown that ϕ is full and isotropic. By explicit construction of an inverse using an interpretation of formulae of Glaser and Stora (13) which involves the concept of associated curves of the harmonic map ϕ, we may show (11):

4. Classification Theorem. *The assignment (5) gives a bijection between the set of full isotropic harmonic maps* $\phi : M \to \mathbb{C}P^n$ *from a Riemann surface to complex projective* n*-space and the set of all pairs* (f,r) *where* $f : M \to \mathbb{C}P^n$ *is a full holomorphic map and* $r \in \{0,\dots,n\}$. *If* M *is compact the degree of* ϕ *is* $\deg f_r - \deg f_{r-1}$.

5. Special Case. Say that a full holomorphic map $f : M \to \mathbb{C}P^n$ is *totally isotropic* if $f_{n-1}^{\perp} = \bar{f}$, equivalently in terms of local lifts f_U of f:

$$\langle \partial_z^{\alpha} f_U, \overline{\partial_z^{\beta} f_U} \rangle = 0 \text{ for all } \alpha,\beta \geqslant 0,\ \alpha + \beta \leqslant n-1.$$

Then n must be even, say, $n = 2r$. And in (2.5), $g = f_{n-1} = \bar{f}$ so

$$\phi(x) = (f_{r-1}(x) \oplus \overline{f_{r-1}(x)})^{\perp},\ x \in M. \tag{6}$$

From this formula we see ϕ has image in $\mathbb{R}P^n \subset \mathbb{C}P^n$ and so we are led to the

6. Classification Theorem (1, 3). *The assignment (6) gives a bijection between the set of all full isotropic harmonic maps* $\phi : M \to \mathbb{R}P^{2r}$ *from a Riemann surface* M *to real projective* 2r*-space and full holomorphic totally isotropic maps* $f : M \to \mathbb{C}P^{2r}$. *There are no full isotropic harmonic maps* $\phi : M \to \mathbb{R}P^n$ *if* n *is odd.*

Remark. By being careful about orientations we can interpret (6) as defining one of two maps into S^{2r}, thus *there is a* 2 : 1 *correspondence between the set of all full isotropic harmonic maps* $\phi : M \to S^{2r}$ *from a Riemann surface* M *to the* 2r*-sphere and full holomorphic totally isotropic maps* $f : M \to \mathbb{C}P^{2r}$. *There are no full harmonic maps* $\phi : M \to S^n$ *if* n *is odd.*

Combining II.4 and II.6, with the isotropy theorems I.4 and I.9 we deduce :

7. Classification Theorem. *The assignment (5) gives a bijection between*

(i) (4, 11) *the set of all full harmonic maps* $\phi : S^2 \to \mathbb{C}P^n$ *and the set of all pairs* (f,r) *where* $f : S^2 \to \mathbb{C}P^n$ *is a full*

holomorphic map and $r \in \{0,\ldots,n\}$;

(ii) (11) *the set of all full harmonic maps* $\phi : T^2 \to \mathbb{C}\mathbb{P}^n$ *of non-zero degree and the set of pairs* (f,r) *where* $f : T \to \mathbb{C}\mathbb{P}^n$ *is a full holomorphic map,* $r \in \{0,\ldots,n\}$ *and* $\deg f_r \neq \deg f_{r-1}$;

(iii) (1,3) *the set of all full harmonic maps* $\phi : S^2 \to \mathbb{R}\mathbb{P}^{2r}$ *and the set of full holomorphic totally isotropic maps* $f : S^2 \to \mathbb{C}\mathbb{P}^{2r}$. (Note full harmonic maps $\phi : S^2 \to S^{2r}$ are in 2:1 correspondence with full harmonic maps $\phi : S^2 \to \mathbb{R}\mathbb{P}^{2r}$ so the former are also classified by this theorem.)

An iterative scheme for finding all full holomorphic totally isotropic maps f is given in (2).

During the preparation of this note the author was supported by the Sonderforschungsbereich "Theoretische Mathematik", Universität Bonn.

REFERENCES

(1) J. Barbosa, On minimal immersions of S^2 into S^{2m}. Trans. Amer. Math. Soc. 210 (1975), 75-106.

(2) H.J. Borchers and W.D. Garber, Local theory of Solutions for the O(2k+1) σ-model, Comm. Math. Phys. 72 (1980), 77-102.

(3) E. Calabi, Quelques applications de l'analyse complexe aux surfaces d'aire minima, Topics in Complex Manifolds, Univ. Montréal (1967), 59-81.

(4) A.M. Din and W.L. Zakrewski, General classical solutions in the CP^{n-1} model. Nucl. Phys. B 174 (1980), 397-406.

(5) J. Eells, Variational theory in fibre bundles: examples, these proceedings.

(6) J. Eells and L. Lemaire, A report on harmonic maps, Bull. London Math. Soc. 10 (1978), 1-68.

(7) J. Eells and L. Lemaire, On the construction of harmonic and holomorphic maps between surfaces, Math. Ann. 252 (1980) 27-52.

(8) J. Eells and J.H. Sampson, Harmonic mappings of Riemannian manifolds, Amer. J. Math. 86 (1964), 109-160.

(9) J. Eells and J.C. Wood, Restrictions on harmonic maps of surfaces, Topology 15 (1976), 263-266.

(10) J. Eells and J.C. Wood, Maps of minimum energy, J. London Math. Soc. 23 (1981), 303-310.

(11) J. Eells and J.C. Wood, Harmonic maps from surfaces to complex projective spaces, Preprint.

(12) A. Futaki, Nonexistence of minimizing harmonic maps from 2-spheres. Proc. Japan Acad. Ser. A Math. Sci. 56 (1980), 291-293.

(13) V. Glaser and R. Stora, Regular solutions of the $\mathbb{C}\mathbb{P}^n$ models and further generalizations, preprint.

(14) R.D. Gulliver, R. Osserman and H.K. Royden, A theory of branched immersions of surfaces, Amer. J. Math. 95 (1973), 750-812.

(15) P. Hartman, On homotopic harmonic maps, Canad. J. Math. 19 (1967), 673-687.

(16) K. Kenmotsu, On minimal immersions of R^2 into S^N, J. Math. Soc. Japan 28 (1976), 182-191.

(17) J.L. Koszul, Lectures on fibre bundles and differential geometry, (Tata Institute, 1960).

(18) L. Lemaire, Applications harmoniques de surfaces riemanniennes, J. Diff. Geom. 13 (1978), 51-78.

(19) L. Lemaire, Harmonic nonholomorphic maps from a surface to a sphere, Proc. Amer. Math. Soc. 71 (1978) 299-304.

(20) A. Lichnerowicz, Applications harmoniques et variétés kählériennes, Symp. Math. III Bologna (1970), 341-402.

(21) B. O'Neill, The fundamental equations of a submersion, Mich. Math. J. 13 (1966), 459-470.

(22) J. Sacks and K. Uhlenbeck, The existence of minimal immersions of the two-sphere, Bull. Amer. math. Soc. 83 (1977), 1033-1036.

(23) R. Schoen and S.T. Yau, Existence of incompressible minimal surfaces and the topology of three dimensional manifolds with nonnegative scalar curvature, Ann. Math. 110 (1979),127-142.

(24) Y.-T. Siu, Some remarks on the complex analyticity of harmonic maps, Southeast Asian Bull. Math. 3 (1979), 240-253.

(25) T. Sunada, Rigidity of certain harmonic mappings, Invent. Math. 51 (1979), 291-307.

(26) J.C. Wood, Harmonic maps and complex analysis, Proc. Summer Course in Complex Analysis, Trieste (1976), vol. III, 289-308.

(27) J.C. Wood, A note on the fundamental group of a manifold of negative curvature, Math. Proc. Cambridge Phil. Soc. 83 (1978), 415-417.

(28) J.C. Wood, Holomorphicity of certain harmonic maps from a surface to complex projective n-space, J. London Math. Soc. 20 (1979), 137-142.

(29) J.C. Wood, On the holomorphicity of harmonic maps from a surface, Global Differential Geometry and Global Analysis, Proceedings, Berlin 1979, (Springer-Verlag, Lecture Notes in Mathematics #838, 1981).

(30) J.C. Wood, Conformality and holomorphicity of certain harmonic maps, preprint.

(31) H.M. Wu, The equidistribution theory of holomorphic curves (Ann. of Math. Studies 64, 1970).

INDEX

GPSR Compliance

The European Union's (EU) General Product Safety Regulation (GPSR) is a set of rules that requires consumer products to be safe and our obligations to ensure this.

If you have any concerns about our products, you can contact us on ProductSafety@springernature.com

In case Publisher is established outside the EU, the EU authorized representative is:

Springer Nature Customer Service Center GmbH
Europaplatz 3
69115 Heidelberg, Germany

Batch number: 10620141

Printed by Printforce, the Netherlands